Air and Gas Drilling
Manual

Air and Gas Drilling Manual

Editor

Suhas Kulkarni

Air and Gas Drilling Manual

Edited by **Suhas Kulkarni**

Printed in 2017

ISBN: 978-1-68117-332-0

Library of Congress Control Number: 2015939244

© 2016 by
SCITUS Academics LLC,
616, Corporate Way, Suite 2, 4766,
Valley Cottage, NY 10989

www.scitusacademics.com

Contents

vi

Preface

Air and Gas Drilling Manual is written as a practical reference for engineers and earth scientists who are engaged in planning and carrying out deep air and gas drilling operations. The book covers air (or gas) drilling fluids, aerated (gasified) drilling fluids, and foam drilling. Further, from the mechanical rock destruction standpoint, the book covers conventional rotary drilling, downhole positive displacement motor (PDM) drilling, and down-the-hole hammer (DTH) drilling. The entire engineering material in both the USCS and SI unit systems, including the important equations used for air (or gas) drilling and aerated (gasified) drilling, along with the foam drilling calculations. Solutions based on these equations are given in MathCad™, using both USCS and SI units.

Editor

Development of a Fuel Efficient Cookstove through a Participatory Bottom-up Approach

Vijay H Honkalaskar[1], Upendra V Bhandarkar[2], and Milind Sohoni[3]

[1]CTARA, Indian Institute of Technology Bombay, Mumbai, 400076, India

[2]Department of Mechanical Engineering, Indian Institute of Technology Bombay, Powai, Mumbai, 400076, India

[3]Department of Computer Science and Engineering, Indian Institute of Technology Bombay, Mumbai, 400076, India

ABSTRACT

Background

Since 1940s, other than a few success stories, the outcomes of efforts of development and dissemination of improved cookstoves [a] have not been so fruitful. This paper presents a bottom-up approach that was successfully implemented to develop a fuel-efficient cookstove in a tribal village that has resulted in a substantial reduction in firewood consumption.

[a]Improved cookstove is generally regarded as a relative concept. Improved cookstoves may improve fuel efficiency, reduce indoor air pollution that causes multiple health hazards, prevent fire hazards, reduce cooking time, or reduce green house gas emissions[56-58].

Method

The approach ensured people's participation at multiple stages of the process that started from project selection by capturing people's needs/desires and studying the existing cooking practice to understand its importance in the local context. The performance of the cookstoves was evaluated by modifying a standard Water Boiling Test to accommodate the existing cooking practice. The improvement of the cookstove was achieved by fabricating a simple twisted tape assembly that could be placed on it without changing the existing cookstove.

Results

The optimization of the twisted tape device was first carried out in the laboratory and then implemented in the field. The field-level tests resulted in reduction of firewood consumption by around 21% which is a substantial improvement for such a device. It was also found that the improvement reduced soot [b] accumulation by around 38% and time of cooking preparations by around 18.5%.

[b]It is a particulate matter especially black carbon.

Conclusion

Overall, a bottom-up and participatory process that not only addressed people's perceived needs but also ensured no changes in the existing cooking practice while providing an easy, low cost (around US$1.25) [c], and locally manufacturable solution led to a highly successful improvement in the local cookstove that was accepted easily by the villagers.

[c]It was assumed that US$1 is around 50 Indian Rupees.

BACKGROUND

The inception of a development discourse in the West around the 1940s [1] has led to development of new technologies to address the problems of the poor in undeveloped and developing countries. Since then, the development and dissemination of improved cookstoves has been one of the key areas for development practitioners in government, volunteer organizations, and research laboratories[2,3]. Other than a few success stories, the outcomes of efforts of these practitioners have not been so fruitful. This section summarizes impacts and challenges of different improved cookstove programs over the past 50 years. Sesan [4] broadly categorized these programs in three phases. We have adopted this categorization in the present context of technology development. These three phases are enumerated below.

Early Programs (1950 To 1980)

Since the 1950s, the development and dissemination of improved cookstoves in the developing countries [3] was, to a large extent, due to initiatives from western countries. According to Crewe [5], fashionable aims such as 'working with poor women', 'raising the income of poor artisans', and 'conserving environmental resources' promoted the involvement of international organizations in the development of improved cookstoves. The main goal of this phase for improved cookstoves was to increase the thermal efficiency of the cooking process to reduce the envisaged energy gap [5] between the supply and demand of biomass fuel due to the thinning of the forests. These early programs assumed that people would adopt the improved stoves quickly and that an initial intervention would lead to a self-sustaining program. It was also assumed that there existed only a need to distribute improved cookstoves, and since these were intrinsically superior to the traditional cookstoves, a decrease in fuel consumption by 75% or even more was imminent [6]. As a consequence of these assumptions, the attempts in this phase failed. The clearest of the lessons learned from early experience is that the chances of success are enhanced when people have an explicit need to save fuel, when the new stoves are a significant improvement over the local traditional stoves, and when stoves can be made readily available by local industries or artisans at affordable prices.

Context Responsive Phase (1980 To 1990)

Researchers started to understand that there is a weak link between deforestation and traditional cookstoves as people cut only small branches to fire their cookstoves [6]. The main objective of the cookstove programs was changed to improve the efficiency of the cookstove to address the perceived needs of users to reduce drudgery involved in the firewood fetching activity. Reduction in indoor air pollution was a secondary objective. Researchers started to emphasize the involvement of local people in the cookstove programs. Yet, these programs faced significant problems in gaining the acceptance of the rural poor [6]. There have been a few partial success stories, mainly in urban areas where fuel has to be bought, and hence, cookstove programs became economical. These include 'Jiko' and 'Upesi' stoves in Kenya [7], Chinese national stoves program [8], 'Anagi' stoves in Sri Lanka, and 'Sara-ole' stove in Karnataka, India [9]. There were two nationwide programs implemented by India and China that followed two different dissemination routes. The Indian National Program on Improved Cookstoves (NPIC) was started in 1983. More than 2.8 million stoves were disseminated until the year 2002 when the government of India officially withdrew funding support from the NPIC. The dissemination of the improved cookstoves was supported by government subsidies. Although it had a high dissemination number, NPIC is not recognized as a successful program because of a very low acceptance rate by users [10-12]. The Chinese National Improved Stoves Programme (NISP), which followed a market-based approach, disseminated 129 million stoves from 1982 to 1992 covering 65% of the Chinese rural population. Most of the cost of stove material and construction labor was paid by the users. However, it must be noted that during this period, NISP operated in relatively middle income areas, and its benefit did not reach the rural poor at that time [8]. The more self-sustaining market-based approach adopted by NISP has led the international government and non-government agencies to follow a market-based approach in the third phase [13].

Third Phase (1990 Onwards): Market-Based Approaches

The focus of improved cookstove programs for this phase has been to reduce indoor air pollution along with increasing cookstove efficiency [4]. Indoor air pollution caused by smoke from traditional cookstoves is responsible for nearly 4.3% of the total global diseases and causes more than 1.6 million deaths each year [14]. Drawing on the experience of the second phase, the market approach was emphasized [13]. For example, in India, there was an emergence of multiple stove designs and dissemination programs funded by private agencies. These involved Shell foundation, Envirofit International, Prakti Design, Selco India, B.P., and TIDE [15,16]. These have resulted in limited penetration, owing to high initial cost and low flexibility towards the use of local fuels [11,16]. Learning from these recent initiatives and outcomes of NPIC, the Indian government started the National Biomass Cookstove Initiative in 2009 [11,17], with the aim of providing energy service comparable to clean sources such as LPG while using the same solid biomass fuel with the combined objectives of fuel efficiency, health protection, and climate impacts. To ensure good quality control and reliable performance, this initiative lays an emphasis on centralized manufacturing facilities. There is narrower tolerance to biomass size and moisture content. Thus, there is an additional requirement of fuel processing at household level.

The Global Alliance for Clean Cookstoves was initiated in year 2010 [17]. Coordinated through United Nations Foundation, it has a goal of providing 100 million clean cookstoves by year 2020 which would reduce the health and other impacts of current cooking practices. In spite of all these efforts, what still remains is to make biomass stoves that are not only clean enough to have major health benefits but also affordable by the poor population of the world.

It can be noticed that most of the approaches adopted for improved cookstoves have remained top down and focused essentially on tangible technical objectives. Many researchers such as Bailis et al. [13] and Barnes et al. [6] acknowledge that although the market-based approach can achieve a sustainable dissemination model, it may filter out the poorer section of rural community. Nevertheless, most of these reports do not analyze the development and dissemination approach in the wider social, cultural, ecological, and economical context in which rural people construct their livelihoods.

Learning From the Improved Cookstove Programs

Goals and strategies to develop and disseminate improved cookstoves have changed over the three phases. These were based on the perceived needs/problems and perceived socio-economic context of community members. There have been a few success stories, where not only have the perceived needs and priorities of the community matched with reality, but this identification has been also followed by sustainable dissemination. Analysis of experiences of a number of (seven) improved cookstove programs around the world revealed the following facts:

- Community participation: It has been widely appreciated that people's participation is very crucial for the success of any development project. After the failure of the first round of improved cookstove projects until 1980, development agencies sought to adopt more participatory ways to develop and disseminate the improved cookstoves [18]. It is considered as vital for facilitating a detailed understanding of user needs and paramount for the success of any development program[19]. Involvement of the community takes care of local manufacture, suitability of improved cookstoves to the local cooking practices, and affordable cost [7,20]. However, even these participatory inputs from the user did not always reflect in a favorable outcome for many projects. The term 'participation' includes notions of people's empowerment, involvement, and control over the development initiative [21-23]. Agencies involved in improved cookstove projects usually have bureaucratic, power-conscious, performance-driven, and goal-oriented structures. It has been very difficult to appreciate people's empowerment and their control in the cookstove development project. Participatory methods adopted by these agencies have been criticized of preserving the participation in terms of mere appearance and actually adhering to a traditional top-down power relation [24]. The Kenyan biomass project in rural areas is an example of such failure of the participatory methods adopted by the Kenyan Ministry of Energy [18] along with donor agencies. Although the improved stove, 'Maendeleo', was very simple and made with locally available technology, not more than 4% of rural households adopted it. Energy saving was not the major concern of the rural dwellers owing to the freely and abundantly available firewood. However, the household income level of the people was very low, and they tended to stick with the traditional practice rather than buying a comparatively expensive (although the stove price was only around US$1.5) alternative. Thus, all the remaining facts can be practically appreciated through people's participation by ensuring their involvement and control in every step.

- User priorities of needs: People tend to adopt interventions on the basis of what matters to them rather than the objectives of the improved cookstove program. The dominance of the interest of outsider stakeholders is one of the reasons of the failures of many improved cookstove programs[18,25]. Improved cookstove projects have been most successful in areas where firewood fetching is an arduous task or people have to spend a significant amount of money on purchasing wood [6]. In both these cases, improved stoves fulfill users' needs. Therefore, the goals of any cookstove program have to be decided on the basis of people's priorities which essentially vary with context.

- Study of existing practices: Cooking practice involves fuel type, fuel size, input power level, number of pots per cookstove, size of pots, time required to cook the food, and ease and flexibility of operation of cookstove. Cooking practices are deeply rooted in the society for hundreds of years[10]. Therefore, an improved cooking method that is different from existing practices has less appeal wherever the people using traditional cookstoves are bound to follow the existing practices because of economic, social, and religious constraints [26]. Therefore, it is imperative to design a new technology that suits the existing cooking practices and also seeks improvements in the performance. One of the reasons behind multiple failures faced by NPIC and a few market-based companies was that the proposed practice of improved stoves could not match the existing cooking practices carried by the community [10]. Notable examples are dissemination of portable stoves in Maharashtra [27] and Philips/BP-Oorja stoves that accommodate only a specific range of fuel type and size [16]. An improved cookstove should be similar to the traditional one to perform the existing cooking practices [7]. Understanding and accommodation of the users' practice to design an improved cookstove are very important for effective adoption and correct utilization of technology [28], even at the expense of not achieving the best efficiency [19,29].

- Other uses: Improved cookstoves do not usually cater to a few other uses and benefits that are sometimes derived from cookstoves such as lighting, space heating, reduction of insects and pests, and drying of thatched roof [30]. The design of improved cookstoves should take into account these characteristics of existing cookstoves.

- Bridging the gap between laboratory and field: One of the most important reasons behind the failure of improved cookstove programs is not meeting the claim to save substantial amounts of fuel. In many cases, improved cookstoves are more efficient than the traditional ones under laboratory conditions, but their performance in field conditions is debatable because improved cookstove designs are found to be incompatible with traditional ways of cooking [30,31]. Oftentimes, the standard Water Boiling Tests (WBTs) are not indicative of the performance of cookstoves in rural communities [32]. This inadequacy can be attributed to lack of stove testing methods which take care of the existing cooking practices [33]. Therefore there, is a need to formulate a performance testing protocol by studying the existing cooking practice to reduce the gap between laboratory- and field-level outcomes.

- Commercialization and dissemination: In many successful programs, the stove dissemination has been an independent entrepreneurial activity. Entrepreneurs generally need technical assistance in designing the stoves and in marketing them to local people [6,29]. A well-designed promotion strategy involves informing the local people about the benefits associated with the improved stoves through a network of local volunteer organizations, television advertisements, or demonstration sites[20]. It is usually found that the local artisans or the trained people seldom adhere to the prescribed dimensions of an improved cookstove [9,16,20]. Thus, for local manufacturing of an improved cookstove, its operation should not be very sensitive to its dimensions.

Overall, it is imperative to understand people's needs and the existing practice to develop an improved cookstove that would be socially and economically acceptable. Improved cookstoves are most popular when they are easily and locally manufactured and have clear advantages in manufacturing cost, fuel economy, matching of existing cookstove practices, ease of use, and durability. Dissemination programs are most effective when they allow for interaction and feedback between stove designers, producers, and users.

Lessons of experiences of early cookstove programs led us to devise a novel approach to address a problem of drudgery involved in firewood fetching activity in a tribal village Gawand wadi (population 293), 120 km from Mumbai, located in Karjat Tribal Block of Raigadh District in the state of Maharashtra in western India (19°04'54.10" N, 73°27'19.34" E). This was achieved by interacting with the villagers over a period of 3 years. The study offers three novel contributions. First of all, it proposes an idea of adapted Water Boiling Test for a given context to measure the performance of a cookstove rather than the usual practice of carrying out standard water boiling tests [34-36]. Many researchers achieve an impressive performance enhancement in improved cookstoves after following a standard WBT in laboratory but fail to obtain the same results on the field [30-33]. The present research shows that the results of Kitchen Performance Tests, which follows field-level practice, are close to the results of the modified WBT carried out in the laboratory. The second contribution is a bottom-up approach for cookstove improvement that takes into account the existing cooking practice and brings a progressive increment in the cookstove performance. This approach increases the chances for the adoption of the improvements as compared to many improved cookstove programs (as discussed in the previous subsections) that build a new cookstove (without taking in to account the existing cooking practice) and try to disseminate it in a particular context. It was proven that the combustion and heat transfer of a cookstove can be improved by the introduction of an optimal combination of twisted tapes (with a particular width, number, and twist angle) in the hearth of the cookstove. It effectively enhances heat transfer coefficient by imparting turbulence to the air draft. At the same time, the turbulence also improves combustion by enhancing mixing of air and fuel.

METHOD

The present method adopts a bottom-up approach that emphasizes on understanding the local context and building on it. Such an approach is credited with having a high potential to result in socially and technologically appropriate solutions. Such solutions, in turn, have a greater probability of widespread adoption and long-term sustainability [18]. Towards this end, people's participation, not only in terms of involvement but also in terms of their control in deciding and planning, was sought in every field-level step that was adopted to develop the fuel-efficient cookstove. These steps involved goal setting, study of existing practice, field-level tests, and feedbacks. An emphasis was placed on the study of existing cooking practices after the appreciation of its importance in the present context. To ensure a high degree of social adoption and so to bear minimal variation from the existing cooking practice, it was decided to make small changes in the existing cookstoves by adopting a few guidelines from existing improved cookstove technologies.

Drawing from the abovementioned bottom-up approach, a different methodology was adopted for successful improvement of the traditional cookstove and its commercialization in Gawand wadi. The steps involved in this methodology are described here in short. Each of these steps is elaborated in later sections.

- Understanding of present context: This was accomplished by studying, documenting, and analyzing the village residents from the view point of their assets, activities, energy resources, and utilization of human work hours to carry out different livelihood activities.

- Goal setting: The goal of this project was decided by involving the villagers thoroughly. People's participation was ensured with the help of a set of surveys/studies. These included the problem-ranking exercise of the Participatory Rural Appraisal (PRA) activity [37], an energy-timeline survey of the domestic activities (similar to the surveys carried by Date [38] and Reddy [39]), and a health survey.

- Field-level investigation of existing practice: This was carried by focused group discussions with women and potters (who make the cookstoves), measurements of existing cookstove dimensions, a walk through the firewood fetching area, and experiments to find the performance of traditional cookstoves. A detailed documentation of the cooking operation was carried out by selecting six households in the village that represent the differences in family size, number of women per unit family, and type of cookstove. The details of the selection procedure are discussed in the section, 'Field-level experimentation'.

- Adapted water boiling tests: Accommodating the context-specific practices, two types of WBT, namely, single-pot and two-pot WBTs, were designed to measure performance indices of the cookstove.

- Improvements in the traditional cookstove: To ensure minimal variation from the traditional practice (as explained in points 3 and 4 of the Section, 'Learning from the improved cookstove programs'), the traditional cookstove itself was sought to be modified by either optimizing the existing design specifications and/or retrofitting a new device (such as a chimney) in the existing cookstove. The modification was first implemented and tested in the laboratory.

- Field-level experimentation: The proposed modification was tested in the field in the six selected families by adopting the modified water boiling tests and kitchen performance test. To mimic the existing cookstove firing practice, the women from the households themselves fired the cookstoves to carry out field-level experiments.

- Local manufacturing and distribution: Local artisans (blacksmiths) were involved in making any device that needed to be added to the cookstove. An arrangement was also worked out so that the blacksmith would make and sell these products along with his regular wares (axe, sickle, etc.).

Women's involvement in terms of documentation, suggestions, and feedback at every stage of the process was the crux of the project.

People's Participation

People in the village have contributed in the following ways: (1) people participated in the identification of the projects based upon their need and desires, (2) women participated in the study of the existing cooking practices, (3) women from six selected households (see Table 1) have participated in conducting field-level experiments, (4) women participated in the modification of the standard water boiling test, and (5) women participated in improving the twisted-tape device to make it user friendly. These persons have consented to take part of this study and the publication of the accompanying images.

Table 1: Selected households to carry out field-level experimentation and survey cooking practices

Households	Family size	Number of women	Number of women per unit family size	Type of cookstove
Ambabai Kisan Gawanda	3	2	0.67	Three-pot stove
Tukibai Dhavlya Gawanda	4	1	0.25	Two-pot stove
Gulab Chander Bhagat	5	1	0.2	Three-pot stove
Hirabai Govinda Gawanda	7	4	0.57	Two-pot stove
Shevanta Dehu Bangare	9	4	0.45	Two-pot stove
Bhimabai Kalu Warghade	14	3	0.21	Three-pot stove

Honkalaskar et al.

Honkalaskar et al. Energy, Sustainability and Society 2013 3:16

doi:10.1186/2192-0567-3-16

Owing to the lack of technical knowhow required to participate in the process of analytical modeling and identification of the retrofitting (twisted-tapes inserts) for the design improvement of the traditional cookstoves followed by its laboratory level design optimization, it was not possible to involve the people in these three activities.

The following section would introduce the context of the study briefly.

Understanding of the Present Context

The major sources of livelihood for the villagers are a single crop rain-fed agriculture (mainly paddy and ragi), collection and selling/consumption of forest produce such as gum, fruits, vegetables, roots, and oil seeds, and agriculture labor in neighboring irrigated area. Occasionally, they also find wage labor work in road construction and infrastructure projects in neighboring areas. The nearest market place is located at a distance of 30 km.

Assets

The average family size is 5.63 with nearly 63% of households between 4 to 8. Percentages of men and women in the population are nearly equal, and the adult population (above 18 years of age) accounts for 71% of the total population. The total village land is 950 ha. Most of the land is sloping and forest land. Total cattle head is 212 with 4.07 cattle head per family, which is higher than the national average of 1.73 [40]. The surrounding forest serves as a source to collect wood fuel, wood, and other forest produce. It comprises both village-owned and government-owned forest land. The main water sources are small earthen dam (around 200 m from the village center, 67,000 m ^{3}capacity) near the village residential area and three wells.

Activities

The villagers mainly depend on local ecological resources for their livelihood, and therefore, their daily activities vary seasonally. They produce most of the things for their consumption. Their daily activities can be listed as follows:

- Rain-fed agricultural activities: These involve Rab preparation[d], ploughing, puddling, transplanting, weeding, harvesting, threshing, and storage. The Rab activity begins in March, ploughing begins in June (following the first monsoon rain), other activities follow in the order mentioned, and storage is in October.

- Livestock raising: Animal raising is mainly done in the monsoon. In other seasons, animals are set free.

- Domestic activities: These involve firewood collection, water fetching, cooking, fish/crab catching, plinth preparation, space heating [e], cloth washing, etc.

- Employment and trade: This involves forest collection, wage labor, forest cutting, carpentry, liquor making, sand collection, bamboo work, moha seeds collection, brick making, etc.

All these activities involve exchange of time and material and energy resources, and some of them involve exchange of money. The main sources of energy are firewood, animal energy, human energy, cattle dung, grass, kerosene, state-supplied grid electricity, and food.

[d]This is a pre-monsoon activity in which grass, tree branches, leaves, and cattle dung are burnt over around one-fifth part of the farm to prepare the land for sowing of paddy.

[e]Space heating is required during winter season to warm the house during night and early morning.

Time Line and Energy Utilization Survey

Different activities in the village are carried out in a varied manner by each household. An activity-time duration survey showed that the contribution of men, women, and children for all activities (mentioned in the previous section) is 34.00%, 62.88%, and 2.94%, respectively. The percentage distribution of human work hours for different activities is as follows: 29.60% (agriculture), 8.40% (animal raising), 48.50% (domestic), and 13.40% (employment and trade). Domestic activities, which demand most of the human work hours, are mainly carried out by women. Firewood fetching activity demands 14% of the total human work hours spent over all the livelihood activities carried out in the village for 1 year. Using the average human energy demands for different livelihood activities [38], the average per capita energy expenditure in each of these activities is listed in Table2. Firewood fetching activity requires 9.67% of total human energy expenditure per year.

Table 2: Average human energy expenditure for different livelihood activities per year

Goal setting

Number	Activities	Annual percapita energyexpenditure (Gcal/year)	Percentage ofenergyexpenditure
1	Animal rearing	15.9	6.43
2	Agriculture	68.0	26.50
3	Water fetching	44.1	17.80
4	Firewood fetching	42.9	9.67
5	Cooking	14.6	8.24
6	Cloth washingand plinthpreparation	9.9	4.03
7	Fishing	3.1	1.25
8	Other activities	64.5	26.04

Honkalaskar et al.

Honkalaskar et al. Energy, Sustainability and Society 2013 3:16 doi:10.1186/2192-0567-3-16

The level of people's participation [41,42] and hence the effectiveness of any technololgy development and dissemination process would increase if the technology intervention addresses a problem that is of high priority to the concerned people. Thus it is imperative to identify the priority that the villagers attach to the need to improve the traditional cookstove.

Perceived Priority of Problems According to Severity

A problem-ranking exercise of PRA was carried out to capture the people's perceptions so that the problems faced by the villagers could be ranked according to their severity. This was achieved by first listing out the problems faced by the villagers followed by individual perceptions about their hierarchy. A total of 37 people including, youth and adult men and women were interviewed. The final hierarchy was found by assigning a weight to each problem equal to the reciprocal of its rank assigned by an individual followed by a weighted average. This study revealed that the firewood fetching activity ranked first among the problems faced by the people. The hierarchy of the problems is shown in Table 3. Presently, both firewood and water fetching activities are carried out by women by carrying loads (around 20 to 32 kg) on their head. Both these activities involve walking on a sloping terrain. These activities have considerable health-related impacts.

Table 3: Overall problem ranking

Problems faced	Young men	Adult men	Adult women	Young women	Overall
Drudgery in firewoodfetching	2	3	1	1	1
Unemployment	1	2	3	7	2
Drudgery in waterfetching	3	5	2	2	3
Lack of healthcare in village	9	1	4	4	4
Low level ofeducation	4	6	8	9	5
Liquor addiction	6	4	7	6	6
Lack of toilets	10	7	5	3	7
Lack of transportationfacilities	5	8	6	5	8
Sporadic electricitysupply	8	10	9	8	9
Snakes bites	7	9	10	10	10

Honkalaskar et al.

Honkalaskar et al. Energy, Sustainability and Society 2013 3:16 doi:10.1186/2192-0567-3-16

Further, 40 women across various age groups were surveyed to identify the health hazards associated with these two laborious activities. The women could be distributed into the following three approximate age groups: women below 30 years (22), between 30 and 50 years (12), and above 50 years (6). The survey revealed that most the women face problems of backache, neckache, calf muscle ache, and fatigue.

These observations combined with the results of the energy and timeline survey and those from the problem-ranking exercise of PRA were shared with the people in a village-level meeting (December 2009). The villagers came to a consensus that the firewood fetching and water fetching activities were the ones involving the most drudgery and should be tackled first.

The present study focuses on the improvement of the efficiency of the traditional cookstove in order to address the most pressing problem of drudgery associated with the firewood fetching activity without undermining its effect on indoor air pollution. The problem associated with drudgery in water fetching was tackled differently.

Cooking Practice

Design of Traditional Cookstove

There are two types of traditional cookstoves, namely, two-pot and three-pot cookstoves, as shown in Figure 1. There is a common fire input port to feed firewood for both of these stoves. There are 26 two-pot stoves and 28 three-pot stoves in the village. Three-pot stoves are usually found in larger families. Women feel that two-pot stoves are more efficient than three-pot stoves. In the case of three-pot stoves, the third hole is used for a short duration of cooking operation during the day. At all other times, flames come out of this third hole, and heat is lost. Yet, as there is an additional hearth, cooking is faster on the three-pot stoves. The average dimensions of the cookstoves in the village are listed in Table 9 in Appendix 1.

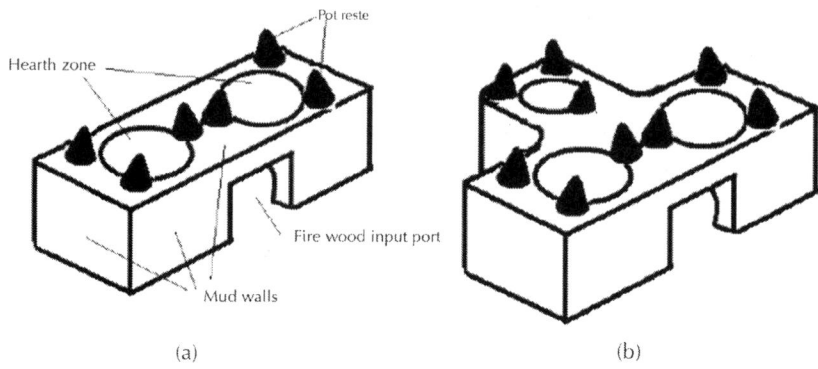

Figure 1: Traditional cookstoves. (a) Two-pot stove and (b) three-pot stove. These stoves are built from clay by people or local potters.

Cookstove Construction

Many households in Gawand wadi make their own cookstove. The remaining households get it manufactured by a few people in the village by offering them a meal. Very few households (presently two) buy cookstoves from potters in nearby villages. The manufacturing process of the cookstoves is elaborated in Appendix 2.

Firewood

The socio-technical aspects of firewood (such as firewood fetching and characterization of firewood) are discussed below:

- Firewood fetching activity: Women fetch firewood from the local forest. Every year, this activity starts after paddy harvesting (November) and continues until the end of the next summer. Figure 2shows an approximate area from which women fetch firewood in Gawand wadi. The dotted line indicates a path travelled by women during winter season to fetch firewood. It is typically 4 to 5 km. The thick line indicates the firewood fetching path during summer season. It is typically 6 to 7 km. The time required to fetch a firewood bundle varies from 3 to 4 h in winter to 5 to 6 h in summer. Initially, tree branches, fallen during monsoon storms, are collected. The area traced per day to collect these branches increases gradually, thereby increasing the daily time required to fetch a firewood bundle. Eventually, when it is very difficult to get fallen wood branches, small-diameter (<5 cm) tree branches are cut. This branch cutting operation demands more human energy than just collection. The typical weight of a firewood bundle carried by women varies from 20 to 32 kg depending upon the load carrying capacity of the woman and the firewood requirement of the household. Figure 3 shows the undulated elevation profile of a path travelled by women to carry firewood during winter. Firewood fetching activity can be divided in to five subtasks. These are to reach the forest area, to collect firewood, to tie the bundle of wood with a thin stem of a creeper or bark of a few tree species, to cut tree branches if required, and to carry firewood home. A typical time breakup required to carry these tasks is shown in Table 4.

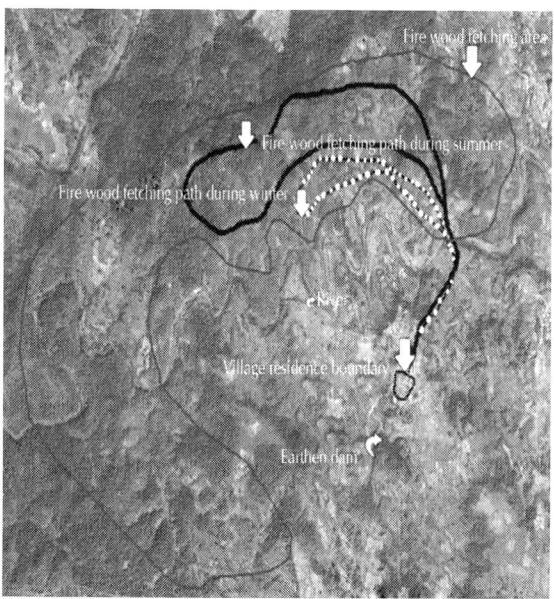

Figure 2: Firewood collection area in Gawand wadi [43]. An approximate area from which women fetch firewood in Gawand wadi. It spans around 10 km². The dotted line indicates a path travelled by women during the winter season to fetch firewood. It is typically 4 to 5 km. The thick line indicates the firewood fetching path during the summer season. It is typically 6 to 7 km.

- Wood species which are used as firewood are ain (Terminalia elliptica), dhamada (Angeissus latifolia), chera (Erinocarpus nimmonii), karvanda (Carissa carandas), bondari (Lagerstroemia parviflora), nana (Lagerstroemia microcarpa), akashi (Salix babylonica), kuda (Holarrhena pubescens), sag (Tectona grandis), etc. Out of these, dhamada, ain, and nana give a good flame. An approximate usage pattern of different firewood species perceived by villagers is shown in Table 5.

- Properties of firewood: Firewood species which are mainly used for cooking were characterized by finding seasonal moisture content, calorific value, and proximate analysis as listed in Table 5. The calorific value was measured using a bomb calorimeter. The proximate analysis was carried using an electric furnace. A dry wood sample (around 1 g) was heated in an electric furnace at 550°C for 7 min and then at 750°C for an additional 2 min to find the percentage of volatile matter in the wood species by measuring the percentage reduction in the weight. It was further heated at 750°C for 4 h to find the percentage of carbon in the sample by calculating the reduction in weight. Tabulated values of calorific values and proximate analysis are averages of three such experiments with less than 5% of variation. Moisture content was measured by the following method: A fixed quantity (100 ± 20 g) of wood was dried in an electric oven at 105°C for 24 h. The fraction of moisture was calculated by subtracting the fraction of dried wood mass. The moisture content varies with season. During rainy season, the moisture content for 3-cm diameter ain wood is around 18% to 20%. It varies from 8% to 14% during summer and winter seasons.

- Typical firewood dimensions: The firewood diameter lies in the range of 1 to 5 cm. To obtain a good flame, firewood of 1.5 to 4 cm diameter is used. For the initial ignition of fire, wood sticks below 1.5-cm diameter are used. Firewood of diameter larger than 5 cm is usually chopped. The length of firewood lies in the range of 50 to 70 cm.

- Wood requirement per year: This varies with family size. For an average family size of six, the wood requirement per year is around 2,700 kg.

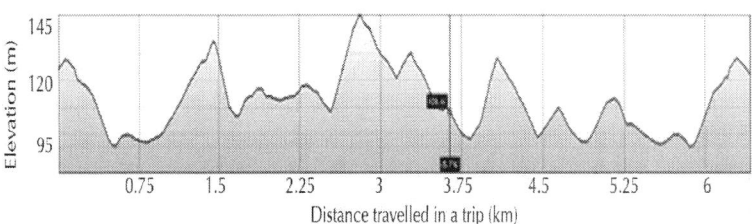

Figure 3: Firewood fetching path during the winter season in Gawand wadi [43]. Elevation of a firewood fetching path traced during winter as shown by dotted line in Figure 5. It shows that the path is quite undulated. The difference between the highest and the lowest location is around 50 m.

Table 4: Approximate time breakup for different tasks performed in firewood fetching activity

Tasks	Average timerequired inwinter (h)	Average timerequired insummer (h)
To reach forest area	0.5	0.7
To collect wood sticks	2	4
To cut tree branches	0	3 (wood eitheris collectedor cut)
To tie bundle with a climber	0.25	0.25
To carry firewood bundle home	0.75	1
Total	3.5	5 to 6

Honkalaskar et al.

Honkalaskar et al. Energy, Sustainability and Society 2013 3:16 doi: 10.1186/2192-0567-3-16

Table 5: Wood species wise usage percentage and calorific value

Wood species	Perceived usage in percentage	Calorific value (MJ/Kg)	Percentage of volatiles	Percentage of carbon
Ain (Terminalia elliptica)	50	16.8	76.5	22.3
Dhamda (Angeissus latifolia)	20	17.9	72	26.5
Chera (Erinocarpus nimmonii)	10	17.9	-	-
Kuda (Holarrhena pubescens)	10	17.3	-	-
Sag (Tectona grandis), karvanda (Carissa carandas), bondari (Lagerstroemia parviflora), akashi (Salix babylonica), and others	10	Sag (16.6), karvanda (16.6),bondari (17.1), Akashi (17.3)	-	-

Honkalaskar et al.

Honkalaskar et al. Energy, Sustainability and Society 2013 3:16 doi: 10.1186/2192-0567-3-16

Cookstove-Firing Practice

This was studied by observation of actual cooking, a questionnaire survey, measurement of coal formation and temperature at different parts of the hearth zone, and direct experience of cookstove firing. The following are the observations:

- Kerosene is used for the initial ignition of firewood. Comparatively dried and smaller diameter wood is preferred for the initial burning. It usually takes 2 to 3 min to achieve a steady-state flame.
- Usually, two or three larger diameter wooden sticks (4 to 6 cm) and a few smaller diameter sticks (2 to 4 cm) are fired simultaneously to attain the required heating rate.
- In general, 50 to 70 cm long wood sticks are preferred so that the fire can be adjusted easily by slight movement of these sticks. Firewood is arranged in such a manner that sufficient primary air can pass through the fuel feed port.

- If necessary, the women blow air by mouth to maintain a good flame size.
- Red hot charcoal helps sustain the flame. It is necessary to remove the excess charcoal if it is accumulated in the hearth zone to make space for the firewood.
- The woman who is cooking sits on one side of the cookstove to avoid exposure to radiated heat from the wood insertion hole.

Durability

On an average, a cookstove lasts for 2 to 3 years before its disintegration. Usually, its fuel feeding hole disintegrates first. It wears due to frequent friction between firewood sticks and its edges. Pot rests also disintegrate as these wear due to regular loading and unloading of pots.

Food Preparations

There are three types of food preparations, namely, baking (bhakari or roasted bread), boiling (heating water, rice, curry, and tea), and frying (vegetables dishes and fish/crab curry). Table 6 lists the average food preparation time, the pot size, and the duration of operation for a two-pot cookstove for a family of six. It shows that nearly 32% of the operations are carried in the single-pot mode, and 68% in the two-pot mode for a two-pot cookstove. Correspondingly, for three-pot stoves, the percentage of operation in the two-pot mode and the three-pot mode is 54 and 14, respectively.

Table 6: Average daily cookstove practice of an average family of size six

Cookstove preparations	Per capita per meal requirement	Duration (min)	Number of daily preparations	Total preparation time (min)	Pot diameter (cm)	Pot height (cm)	Single-pot mode (min)	Two-pot mode (min)	Wood consumption (kg)
Water boiling for bathing	4.2 l	21±2	5	105	27.5 ± 1.5	16 ±2	42	63	2.9
Tea	76.3 g	10±1	3	30	14±1	7±1	10	20	0.75
Rice	141 g	24±4	2	50	19±2	12±2		48	1.3
Bhakri	162 g	24±10	Once in 3 days	8	26±2	5		8	0.2
Dal[a]	24 g	35±5	2	80	17±2	9±2		70	1.7
Bhaji		12±3	2	24	26±2	5		24	0.62
Total							55	115	7.4

[a]Here, dal is referred to as *tur dal* (yellow pigeon pea) and *kala ghevda* (a Maharashtrian bean).

[a]Here, dal is referred to as tur dal (yellow pigeon pea) and kala ghevda (a Maharashtrian bean).

Honkalaskar et al.

Honkalaskar et al. Energy, Sustainability and Society 2013 3:16 doi: 10.1186/2192-0567-3-16

Input Power of Cookstove

The input power is the energy input to the cookstove per unit time. It varies with recipes, cookstove dimensions, family size, and cookstove firing practice. The capacity of a cookstove increases with the size of cookstove or the volume of hearth zone. Usually, tea brewing and rice making requires lesser amount of input power. Other practices such as water boiling require higher amount of power input to the cookstove. If the number of women in the family is low (one or two), they tend to reduce the time required for cooking by firing the cookstove at higher input power in order to spare more time for other activities.

Power input to the cookstove is evaluated by two different ways as follows:

- The ratio of hearth zone volume to input power of cookstove follows a rough rule of thumb of 0.6 l/kW [44]. Using an average hearth zone volume of 9.4 l (for two-pot cookstove), the power input to the cookstove is 15.6 kW if it is used in the two-pot mode. For single-pot mode operation, the average hearth zone volume is 6.5 l, and thus, the estimated power input is 10.8 kW.

- Water boiling tests carried out on the selected cookstoves yielded an average input power of around 14.6 kW for two-pot mode of operation and 8.46 kW for single-pot mode of operation. This value is closer to the input power (14 kW) measured by Geller [45] in the village of Ungra near Bangalore.

Experimental Procedure

Table 6 shows that around 40% of the daily firewood is utilized in water boiling for bathing. It is the operation consuming the most amount of firewood. Other activities including tea brewing and rice and curry making are close to the water boiling operation, and these activities cumulatively comprise more than 85% of the cooking operations. Therefore, a WBT and a kitchen performance test (KPT)[36,46,47] that involves a real time evaluation of the daily cooking practice were used to the water boiling test, a fixed amount of water is boiled on a cookstove, and the efficiency is calculated by dividing the total heat transferred to water by the total heat input.

The components of the cookstove system are operator, firewood, cookstove design, pot, and type and amount of food to be cooked. Each component plays a role to influence the performance of a cookstove. Therefore, the water boiling test was modified to accommodate existing cooking practice defined by these components. The modifications are listed below:

- To mimic existing cookstove firing practice, it was decided to let women fire the cookstove while carrying WBTs on field (see Figure 4).
- Time duration of WBT: Most of the cooking preparations in the village last for a little less than half an hour as documented in Figure 5. Therefore, it was decided to carry WBT for half an hour.
- Fuel size: For field-level testing, wood of 2 to 5 cm diameter and 50 to 70 cm length was used, the wood species being mainly ain and dhamda. For laboratory level testing, wood of 2.5 ± 0.5 cm diameter and 25 ± 2 cm length was used.
- Power input: 8.5 kW (1.81 kg firewood/h) in single-pot mode and 14 kW (3.2 kg firewood/h) in two-pot mode of operation.
- Weight of water: 5.5 l (average water usage for the cooking preparations).

- Pot size: 26 to 31 cm (range of pot sizes used in the village to boil the bathing water).
- Timing to carry out tests: 11 am to 5 pm.
- Starting condition: Hot start, owing to the hot start condition for most of the cooking preparations. This step was added after obtaining feedback from the women.

Figure 4: A woman firing cookstove in a two-pot water boiling test. Water boiling tests in the village were carried out with women, where a woman who cooks food fired her cookstove to find out its performance.

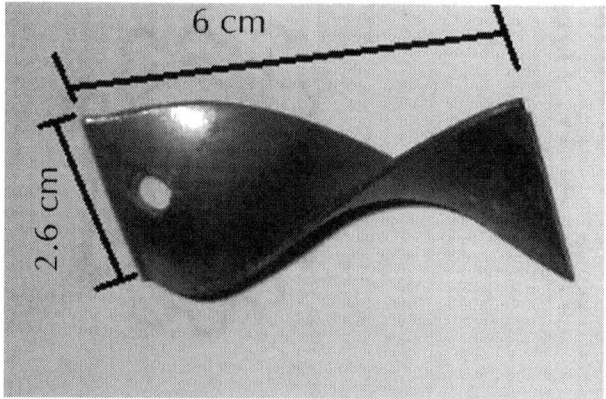

Figure 5: A twisted mild steel plate.

Kitchen Performance Test

KPT measures the average rate of firewood consumption by a stove in a normal household environment [36]. A wood pile (mainly including the wood species ain and dhamda) of 100 ± 15 kg was stacked in a household, and the rate of consumption of firewood per household was found by dividing the total consumed weight of wood by the number of days of its consumption.

Proposed Improvement

Cookstove operation comprises three phenomena that occur simultaneously, namely, (a) burning of fuel, (b) mass transfer of air, volatile gases, and gaseous products of combustion, and (c) heat transfer to pot and other parts of the stove [48,49]. The overall thermal efficiency (ηt) of the cookstove depends upon three intermediate efficiencies, namely, combustion efficiency of fuel (ηc), heat transfer efficiency to pot (ηh), and efficiency at which pot transfers energy to food (ηp). The efficiency value ηp depends upon pot size which is usually determined by food type. The thermal efficiency can be improved by either optimizing the design specifications (see Figure 12 in Appendix 1) or retrofitting a certain device (such as a chimney) in the existing cookstove or both to improve heat transfer efficiency and/ or combustion efficiency.

To study the cookstove operation for its design improvement, a steady state analytical model was developed by splitting the hearth into three zones to study char combustion, volatile combustion, and heat transfer to the pot bottom separately. A study of the variation of design specifications on the thermal performance revealed that the traditional cookstove specifications are close to their optimal values. A Sankey diagram (see Figure 15 in Appendix 3) of the cookstove operation revealed that the major thrust areas to improve the thermal performance are the heat transfer to the pot and combustion of volatiles[f].

It has been documented that the incorporation of twisted tapes (TTs, see Figure 5) in the hearth generate swirl motion of the gases which improves the jet impingement heat transfer [50,51] at the pot bottom. It also improves combustion of volatiles by increasing air-fuel mixing and the residence time of the reactant gases [52,53]. Bhandari et al.

[47] found that the inclusion of twisted tapes improved the efficiency of an experimental cookstove. It was decided that the effect of the inclusion of twisted tapes to improve the thermal performance of the traditional cookstove be investigated.

ᶠThe details of the thermo-chemical model of the traditional cookstove are not discussed in the present report and will be presented elsewhere.

RESULTS AND DISCUSSION

Effect of Variation of Number of Twisted Tapes on Thermal Performance of Cookstove

Manufacturing in the Laboratory

The twisted tapes were manufactured by twisting a mild steel plate heated to 300°C. The strip dimensions used for laboratory tests were the following: for the width, 1.25, 2.54, and 3.38 cm; length, 6 cm; and thickness, 1.5 mm. The tapes were twisted by holding them in a fixture together with a lathe machine. To ensure smooth twist, the tapes were heated with a torch blower, and the twisting procedure was implemented in four to five steps.

Retrofitting of Twisted Tapes

TTs were retrofitted in a cookstove by just inserting them in two or more steel rods and keeping the rods on the holes of the hearth to let the TTs hang in the hearth as shown in Figure 6. Thus, the intervention does not change the existing cookstove design. The new cookstove is similar to the existing one and would address the concerns raised in point numbers 3 and 4 of the Section "Learning from the improved cookstove programs".

Figure 6: Twisted tapes hanging in the hearth holes. This shows the way to hang the twisted tapes in the hearth holes by using steel rods. They are placed in a way that they cover the cross-section of the hearth evenly.

Outcomes of Laboratory Tests

A two-pot cookstove, made by a local potter, was used for the laboratory-level experimentation. Its specifications are listed in Table 10 in Appendix 1. The two-pot water boiling tests were carried out by varying the number of TTs in each hearth (experimental set-up is shown in Figure 7). The water boiling tests were carried out in the following order:

- Varying the number and width of the twisted tapes in each hearth for the same twist angle (180°).
- Varying the twist angle (0°, 60°, 120°, 180°, and 240°) for optimum width and number of twisted tapes (2.54 cm and seven, respectively).
- Carrying a few additional WBTs with optimum width and twist angle (2.54 cm and 180°, respectively) to study the effect on exhaust gas composition and soot. This helps in understanding the contributions due to the twisted tape assembly.

Figure 7: Experimental set-up to carry two-pot WBT in the laboratory. This shows the experimental set-up that was used to carry out the laboratory WBTs. Four thermocouples were used to measure the temperature of water and flame for both the pots and hearths.

It was observed that the thermal efficiency of the cookstove was optimum (improvement by 24.5%) for inclusion of seven number of twisted tapes having a width of 2.54 cm and a twist of 180°. A few additional WBTs with optimum width and twist angle showed that the proposed retrofitting improves combustion of the volatile gases (see Table 7). It was also observed that the amount of soot accumulated over the outer surface of the pot decreases by around 38% (see Table 7) by inclusion of these TT inserts. A detailed analysis of the laboratory level WBTs is not discussed in the present report and will be presented elsewhere. It was decided that field-level tests by retrofitting seven TTs of 2.54 cm width and 180° twist angle in each hearth be carried out.

Table 7: Soot accumulation and percentage composition of the gases at the center of the stove top

	$O_2(\%)$	$\dfrac{CO}{CO_2}$	Soot (g)
Normal cookstove	10.5	0.087	2.84
Seven TTs	5.5	0.082	1.75
Eleven TTs	7.8	0.080	2.20

The listed values are averages of three WBTs.

Honkalaskar et al.

Honkalaskar et al. Energy, Sustainability and Society 2013 3:16 doi:10.1186/2192-0567-3-16

Field Level Experimentation

Field-level experiments were carried out at the six selected households in the following order: single-pot water boiling test, two-pot water boiling test, and kitchen performance test. The method of selecting the households is explained below.

Selection of the Households

It was first noted that the cookstove size and firewood requirement increase with family size (see Figures 13 and 14 in Appendix 1). If the number of women per unit family size is higher, firewood fetching and cooking activities get distributed among women, and it reflects in the human work hour involvement and drudgery associated with it. If the number of women per unit family size is lower, women try to fire the cookstove with a higher input power in order to accomplish the cooking task earlier. Figure 8 shows all the existing combinations of the family size and number of women per unit family size in the village. The data points are then divided in four quadrants as shown. Depending on the willingness of the family members to participate in field-level experimentation and the type of cookstove (two-pot or three-pot), four different households from each quadrant were selected along with two households from extreme ends (encircled in Figure 9 and listed in Table 1). As there are nearly equal number of two-pot and three-pot stoves in the village, it is ensured that three out of the six selected households have three-pot stoves. The dimensions of the cookstoves are listed in Table11 in Appendix 1.

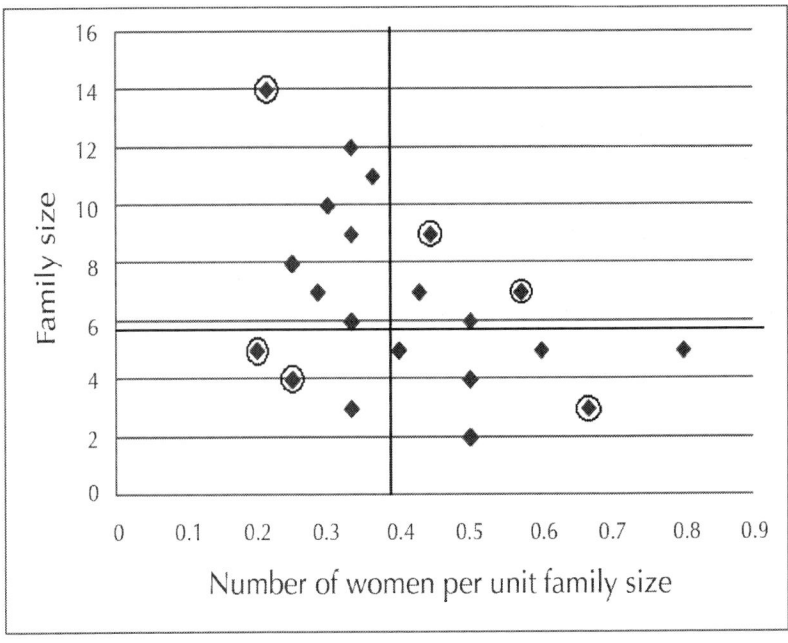

Figure 8: Family size versus number of women per unit family wise in Gawand wadi. The graph shows all possible combinations of family size and women per unit family size. The sample space is divided in four quadrants for even selection of the households across such variations. Encircled families were selected.

Figure 9: TT pack. This shows a twisted tape pack that can be retrofitted in the traditional cookstove.

Single-Pot Water Boiling Test

By introducing seven TTs, the thermal efficiency of the cookstove increases by an average of 20.61 ± 1.98%, and the specific fuel consumption of cookstove is decreased by 14.5 ± 1.32% (at the statistical significance level of 0.05 for both the outcomes).

Two-Pot Water Boiling Test

The thermal efficiency of the cookstove increases by an average of 22.82 ± 2.48%, and specific fuel consumption of the cookstove decreases by 15 ± 1.29% (at the statistical significance level of 0.05 for both the outcomes) by retrofitting of TTs in existing cookstove (see Figure 10 and 11).

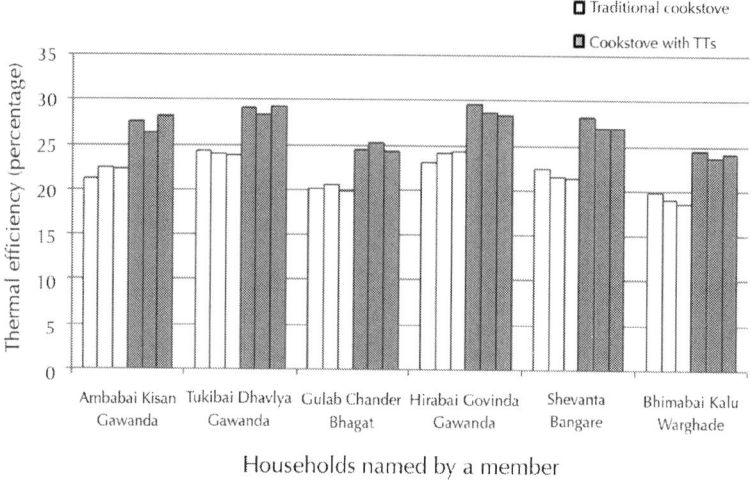

Figure 10: Effect of retrofitting of TTs on thermal efficiency of traditional cook-stoves in Gawand wadi. This shows the results of the two-pot water boiling tests that were carried out in six selected households. Traditional cookstove and cookstove with inclusion of TTs were tested by carrying out three tests for each kind of cookstove. It can be observed that the thermal efficiency of the cookstove increases by the inclusion of TTs for every household.

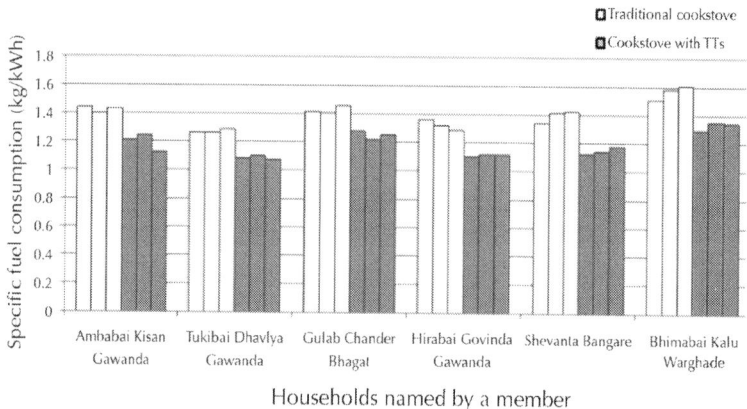

Figure 11: Effect of retrofitting of TTs on specific fuel consumption of traditional cookstoves in Gawand wadi. Figure shows results of the two-pot water boiling tests that were carried out in six selected households. Traditional

cookstove and cookstove with the inclusion of TTs were tested by carrying out three tests for each kind of cookstove. It can be observed that the specific fuel consumption of the traditional cookstove decreases by the inclusion of TTs for every household.

Kitchen Performance Test

The results of the KPT are shown in Table 8. Corrections in wood consumption of the families due to presence of guests were incorporated by considering an average firewood usage for food preparations as listed in Table 6. The results of KPT show (see Table 8) a reduction of 21.3 ± 1.89% (at the statistical significance level of 0.05) in the average daily wood consumption.

Table 8: Results of kitchen performance test at two households

Household	Operation conditions	Average daily wood requirement (kg/day)
Ambabai Kisan Gawanda	Normal cookstove	5.81
	Cookstove with TTs	4.56
Tukibai Dhavlya Gawanda	Normal cookstove	5.78
	Cookstove with TTs	4.39
Gulab Chander Bhagat	Normal cookstove	7.25
	Cookstove with TTs	5.56
Hirabai Govinda Gawanda	Normal cookstove	7.22
	Cookstove with TTs	5.89
Shevanta Dehu Bangare	Normal cookstove	9.08
	Cookstove with TTs	7.43

Bhimabai Kalu Warghade	Normal cookstove	9.39
	Cookstove with TTs	7.38

Honkalaskar et al.

Honkalaskar et al. Energy, Sustainability and Society 2013 3:16 doi:10.1186/2192-0567-3-16

The results of the kitchen performance test show that the percentage increase in the performance of cookstove is lower than that indicated by WBTs. Once the cooking preparations in the morning, noon, or evening are completed, the women leave accumulated coal inside the hearth itself. It slowly burns and decreases in weight until the next operation starts. Secondly, if the coal accumulation prohibits insertion of firewood in the hearth zone, the women take the coal out of the cookstove and throw it away. Thus, all the coal that remains at the end of cooking operations is not utilized. Hence, the percentage decrease in wood consumption is lower (by around 3%) than the percentage increase in the thermal efficiency. Yet, as some amount of the remaining coal is still utilized for further food preparation, the percentage decrease in average wood consumption is higher in KPT than the percentage decrease in specific fuel consumption (by around 4%) in WBT.

As KPT gives actual improvements in the cookstove performance by enabling real-time monitoring of the cooking performance, it is a more relevant test. Furthermore, as water boiling is a major cookstove operation (see Table 6), the results shown by the water boiling tests are closer to those shown by the kitchen performance test.

Women's Feedback

The women in the six selected households gave the following feedbacks regarding the retrofitting of TTs in their cookstoves:

- Improvement in the cookstove has reduced wood consumption by roughly 25% (as per their estimate).
- Soot accumulation over the outer surface of pots has reduced.
- Time required to prepare different food items has reduced.

- Women found that sometimes the steel rod holding the TT would fall inside the hearth zone due to lateral force exerted by firewood on TTs. If this problem is removed, it would be wonderful.

A new retrofitting was devised as a result, in which the seven TTs are inserted in a welded triangular rod structure as shown in Figure 9. It does not fall in the hearth and is very easy to place and remove from the stove. Women liked this new design.

Commercialization through Local Artisans

Commercialization involves manufacturing and distribution of the technology. Bottom-up approach emphasizes building on the local human skills and the local socio-economic structure. The improvement is found to be not very sensitive to the slight changes in the width and twist angle of the twisted tape. Thus, local manufacturing and dissemination of the TT packs was found to be an impressive alternative. The success of the dissemination of an improved cookstove program depends upon multiple factors [54] that vary with context. Although many dissemination programs emphasized on training local people for manufacturing and marketing of improved cookstoves[20,55], it was found that at many places, these enterprises get poor benefits, so they have little motivation to start and sustain the business [9]. Therefore, if a local artisan is trained to accommodate the manufacturing and selling of improved cookstoves in his/her existing socio-commercial business, there is a greater probability for its sustenance. In the present context, manufacture and dissemination of the TT packs through the existing business structure of the local blacksmiths is sought. Traditionally, they make and repair sickles, axes, ploughs, and other tools required for agricultural and domestic activities. Recently, many of them have diversified their skills to lay tin roofs and build steel gates. Manufacturing of TT pack involves cutting, drilling, heating, twisting, and welding operations. Thus, the local blacksmiths who have welding and drilling machines can manufacture the TT pack. There is a traditional custom of buying and repairing the iron/steel tools from local blacksmiths. Thus, the manufacture and selling of the TT packs can be accommodated in the existing sustainable traditional business of the blacksmiths.

A blacksmith's work load is seasonal. He makes and repairs ploughs and sickles during monsoon for agricultural activities, while he makes axes and other tools in the winter season. During summer, he is engaged in house building activities that involve welding, tin/steel cutting, and drilling operations. Usually, a blacksmith has leisure time during monsoon season. Thus, the blacksmiths can allot more time to make TT packs during monsoon.

A manufacturing process to build the TT packs has been devised by interacting with two blacksmiths and one steel fabricator [g].

- A steel plate with thickness of 1.5 to 2 mm is cut to get small plates of required TT size.
- The plates are placed in the blacksmith's kiln to make them red hot.
- The red hot plate is held in a vise and twisted by a long lever with a groove to hold the plate. It is twisted by 60° to 90° in one go. It is then again placed in the kiln to heat. The whole process of twisting it to 180° requires two to three steps. This is to ensure a smooth curve for the twisted tape.
- A hole is drilled in the TT to insert a steel rod (4 mm diameter).
- Finally, a TT pack is made by welding the rods together.

The present material cost is US$0.5 per TT pack. By considering the manufacturing cost, transport cost to buy steel plate and rods, and the profit of blacksmith (US$8 per day), the total cost comes to around US$1.25 per TT pack. The technology is being disseminated with the help of two local volunteer organizations, Disha Kendra and Oak Vanaushadhi Kendra.

[g]Madhukar Chavan from Kashele, Anil Jadhav from Kondivade, and Pankaj Gore from Vaingaon.

Dissemination of the Innovation

As of now, 45 families from 11 different villages [h] have been using the TT packs in their cookstoves. Ten out of these families have been using this device for more than a year. Similar traditional cookstoves are found across the coastal belt of Maharashtra. Thus, the same innovation can be disseminated across the villages in this coastal region. For the other parts of rural India, the stove designs are much different, and there is a need to identify a context-specific intervention, which may not necessarily be a twisted tape.

[h]These villages are Gawand wadi, Tepachi wadi, Waghai wadi, Kothimbe, Rajpe, Kalamb, Vare, Kondivade, Tamhanath wadi, Khandpa, and Dhare wadi located in Karjat block of Raigadh district of Maharashtra state.

CONCLUSIONS

The present study derives a novel bottom-up approach to develop or improve a rural/community technology by appreciation of wider aspects of the technology-community linkage, namely, people's priorities, people's desires behind improvement or development of technology, importance of studying and maintaining existing technology practises, availability of local resources to manufacture and operate the technology, and its commercial viability and sustainability. It enhances social efficiency of adoption of the technology by people. Overall, the approach mingles three aspects linked with rural/community technology.

- Social: This includes people's participation to set the goal behind the development of technology, appreciation of the existing context in terms of livelihood activities, local natural resources, and human skills.
- Technical: This includes the formulation of a procedure to measure the performance of the technology according to existing context, laboratory and field-level testing, feedback from the people, and further improvements in the technology.
- Commercial: Manufacturing and distribution through locally sustainable structure.

Retrofitting of the TT pack in the existing cookstoves in the village would save nearly 22% of human work hours and human energy involved in the firewood fetching activity. For an average family of six, women make 97 to 122 trips per year to the local forest to fetch firewood. These trips can be decreased by 21 to 27. By considering an average duration of 4.5 h/trip, there can be an annual saving of 95 to 122 women work hours. The TT pack also decreases soot accumulation on the outer surface of pot by around 38%. Based on the temperature variation graph of the water boiling test in the laboratory, the TT pack reduces the cooking time by around 18.5%. For an average family of six, it saves 0.5 h daily and 182 h annually required for cooking preparations. Although the laboratory experiments showed that the optimal retrofitting of the twisted tapes improves combustion and thereby reduces indoor air pollution, it remains to measure the percentage reduction of health-damaging emissions [i].

[i]Major health-damaging emissions involve particulate matter, carbon monoxide, carcinogens such as benzo[a]pyrene, formaldehyde, nitrogen dioxide, and sulfur dioxide[59,60].

AUTHORS' CONTRIBUTIONS

VHH contributed in capacity development, carrying out field-level investigations, field tests, and report writing. He is thus the lead contributor to the paper. UVB and MS have contributed in providing guidelines to devise the methodology and providing inputs on the experiments and in the overall sense to structure the report. All authors read and approved the final manuscript.

Acknowledgements

The authors are thankful to the people of the village Gawand wadi, especially to the women, for their participation and support throughout the research work.

REFERENCES

1.   Brown E (1996) Deconstructing development: alternative perspectives on the history of an idea. J Hist Geogr 22(3):333-339

2.   Smith KR (1989) Dialectics of improved stoves. Econ Polit Weekly 24(10):517-522

3.   Westhoff B (1995) One of the oldest technologies in the world - from the open hearth to the microwave. In: Westhoff B, Germann D (eds) Stove images: a documentation of improved and traditional stoves in Africa, Asia and Latin America 1995, Frankfurt: Brandes and Apsel.

4. Sesan TA (2011) What's cooking? Participatory and market approaches to stove development in Nigeria and Kenya. University of Nottingham: Doctoral thesis.

5. Crewe E (1997) The silent traditions of developing cooks. In: Grillo R, Stirrat R (eds) Discourses of development: anthropological perspectives, Oxford: Berg.

6. Barnes DF, Openshaw K, Smith KR (1994) What makes people cook with improved biomass stoves? A comparative international review of stove programs. Washington DC: The World Bank: Technical report, World Bank Technical Paper Number 242, Energy Series.

7. Sanchez T (2008) Lessons from project implementation on cook stoves and rural electrification, the practical action experience. Rugby: Technical report, Practical Action.

8. Smith KR, Shuhua G, Kun H, Daxiong Q (1993) One hundred million improved cookstoves in China: How was it done? World Dev 21(6):941-961

9. Jagdish KS (2004) The development and dissemination of efficient domestic cook stoves and other devices in Karnataka. Curr Sci 87(7):926-931

10. Krishna C (2000) Improved cook stoves; yet to be a success story.http://bioenergylists.org/files/IMPROVED_COOK_Stoves_Evaluation.pdf Accessed 21 Feb 2012

11. Venkataraman C, Sagar A, Habib G, C NL, Smith K (2010) The Indian National Initiative for Advanced Biomass Cookstoves: the benefits of clean combustion. Energ Sus Dev 14:63-72

12. Kishore V, Ramana P (2002) Improved cookstoves in rural India: how improved are they? A critique of the perceived benefits from the National Programme on Improved Chulhas (NPIC). Energ 27:47-63

13. Bailis R, Cowan A, Berrueta V, Masera O (2009) Arresting the killer in the kitchen: the promises and pitfalls of commercializing improved cookstoves. World Dev 37(10):1694-1705

14. Lim SS, Vos T, Flaxman AD, Danaei G (2013) A comparative risk assessment of burden of disease and injury attributable to 67 risk factors and risk factor clusters in 21 regions, 1990Ū2010: a systematic analysis for the global burden of disease study 2010. Lancet 380(9859):2224-2260

15. Rai K, McDonald J (2009) Cookstoves and markets: experiences, successes and opportunities. London: Technical report, GVEP International.

16. Dhoble R, Bairiganjan S (2009) Cooking practices and cook stoves field insights: a pilot study of user experience with traditional and improved cook stoves. Technical report, Centre for Development Finance, Institute for Financial Management and Research. India: Chennai.

17. Smith KR (2010) What's cooking? A brief update. Energ Sus Dev 14:251-252

18. Sesan TA (2012) Whose priorities? Evaluating objectives of participatory development in the context of household energy projects in Africa.http://www.gbengasesan.com/temidocs/WhosePriorities.pdf . Accessed 23 Oct 2012

19. Rouse J (1999) Improved biomass cookstove programs: fundamental criteria for success. The University of Sussex: Masters thesis, The Centre for the Comparative Study of Culture, Development & the Environment.

20. Ergeneman A (2003) Dissemination of improved coosktoves in rural areas of developing world: recommendations for the Eritrea dissemination of improved stoves program. Eritrea: Technical report, Eritrea Energy Research and Training Center.

21. Oakley P (1998) Strengthening people's participation in rural development. New Delhi, India: Society for Participatory Research in Asia, Occasional Paper Series No. 1.

22. Pearse A, Stiefel M (1980) Inquiry into participation: a research approach. Geneva: Technical report, United Nations Research Institute for Social Development.

23. Westergaard K (1986) People's participation, local government and rural development: the case of West Bengal, India. Copenhagen: CDR research report 8, Centre for Development Research.

24. Parfitt T (2004) The ambiguity of participation: a qualified defence of participatory development. Third World Q 25(3):537-555

25. Troncoso K, Castillo A, Masera O, Merino L (2007) Social perceptions about a technological innovation for fuelwood cooking: case study in Rural Mexico. Energ Policy 35:2799-2811

26. Shrestha SK, Thapa R, Bajracharya K (2003) National improved cook stove dissemination in the mid-hills of Nepal, experiences, opportunities and lesson learnt. Yogyakarta, Indonesia: Technical report, The Asia Regional Cookstove Program Network.

27. Sinha B (2002) The Indian stove programme: an insiders' view - the role of society, politics, economics and education. Boiling Point 48:23-26

28. Ruiz-Mercado I, Masera O, Zamora H, Smith KR (2011) Adoption and sustained use of improved cookstoves. Energ Policy 39:7557-7566

29. Rouse J (2002) Community participation in household energy programmes: a case-study from India. Energ Sus Dev 6(2):28-36

30. Quadir SA, Mathur S, Kandpal TC (1995) Barriers to dissemination of renewable energy technologies for cooking. Energ Policy 36(12):1129-1132

31. Tata Energy Research Institute (1989) Evaluation of improved stoves in Tamil Nadu, Rajasthan and West Bengal. New Delhi: Technical report, Submitted to the Department of Non-conventional Energy Sources, New Delhi, Tata Energy Research Institute.

32. Berruetaa VM, Edwardsb RD, Maserac OR (2008) Energy performance of wood-burning cookstoves in Michoacan, Mexico. Renew Energ 33:859-870

33. Makonese T, Chikowore G, Annegarn HJ (2011) Potential and prospects of improved cookstoves (ICS) in Zimbabwe. Cape Town, 11–13 April 2011: Paper presented at domestic use of energy (DUE) conference.

34. Bailis R, Ogle D, MacCarty N, Still D (2007) The water boiling test. Technical report, Shell Foundation. http://ehs.sph.berkeley.edu/hem/content/WBT_Version_3.0_Jan2007a.pdf

35. Bailis R, Ogle D, MacCarty N, Still D (2004) The Water Boiling Test (WBT), version 1.5. Technical report, University of California-Berkeley.

36. Volunteers in Technical Assistance International Standards (1985) Testing the efficiency of wood burning cook stoves. Virginia, USA: Technical report, VITA.

37. Chambers R (1994) The origins and practice of participatory rural appraisal. World Dev 22(7):953-969

38. Date AW (1989) Energy utilization pattern of Shilarwadi. Indian J Rural Technol 1:33-63

39. Reddy AKN (1982) Rural energy consumption patterns - a field study. Biomass 2:255-280

40. National Sample Survey Organization (2010) Press note on income, expenditure and productive assets of farmer households (January–December 2003).http://pib.nic.in/release/rel_print_page1.asp?relid=14531 . Accessed 2 Jan 2012

41. Arnstein SR (1969) A ladder of citizen participation. J Am I Plann 35(4):216-224

42. Aref F, Redzuan M (2009) Assessing the level of community Participation as a component of community capacity building for tourism development. Eur J Soc Sci 8:68-75

43. Google (2012) Maps.http://maps.google.com/ Accessed 20 March 2012

44. Lepeleire GD, Krishna Prasad K, Verhaart P, Visser P (1981) A woodstove compendium. Netherlands: Technical report, Wood burning stove group, Eindhoven University of Technology.

45. Geller HS (1982) Cooking in the Ungra area: fuel efficiency, energy losses, and opportunities for reducing firewood consumption. Biomass 2(6):83-101

46. Bhatt MS (1982) The efficiencies of firewood devices (open fire stoves, chulhas, and heaters). Indian Acad Sci 5(part 4):327-342

47. Bhandari S, Gopi S, Date AW (1988) Investigation of CTARA wood-burning stove. Part 1. Experimental investigation. Sadhana 13(4):271-293

48. Baldwin SF (1952) Biomass stoves: engineering design, development, and dissemination. 1600 Wilson Boulevard, Suite 500, Arlington, Virginia, 22209, USA: Technical report, VITA.

49. Food and Agriculture Organization of The United Nations (1993) Improved solid biomass burning cookstoves: a development manual. Bangkok: Technical report, Regional Wood Energy Development Programme In Asia, GCP/RAS/154/NET, Field Document No.44.

50. Nuntadusit C, Waehahyee M, Bunyajitradulya A, Eiamsa-ard S (2012) Heat transfer enhancement by multiple swirling impinging jets with twisted-tape. Int Commun Heat Mass 39:102-107

51. Kinsella C, Donnelly B, O'Donovan TS, Murray DB (2008) Heat transfer enhancement from a horizontal surface by impinging swirl jets. Paper presented at the 5th European thermal-sciences conference, The Netherlands, 18–22 May 2008.

52. Guillaume DW, LaRue JC (1995) Combustion enhancement using induced swirl. Exp Fluids 20:59-60

53. Benajes J, Molina S, Garcya JM, Riesco JM (2004) The effect of swirl on combustion and exhaust emissions in heavy-duty diesel engines. J. Automobile Eng 218:1141-1148

54. Ramirez S, Dwivedi P, Bailis R, Ghilardi A (2012) Perceptions of stakeholders about nontraditional cookstoves in Honduras. Environ Res Lett 7(4):2-11

55. Karve P (2005) A model for dissemination of improved biomass fuels and cooking devices through rural enterprises. Boiling Point 50:26-28

56. The World Bank (2011) Household cookstoves, environment, health, and climate change: a new look at an old problem. Washington, DC: Technical report, The World Bank.

57. A Volunteer Group of Foresters in Nepal (2013) Improved cooking stoves.http://www.forestrynepal.org/wiki/117 webcite. Accessed 8 May 2013

58. Wikipedia (2013) Cook stove.http://en.wikipedia.org/wiki/Cook_stove webcite. Accessed 8 May 2013

59. Bruce N, Perez-Padilla R, Albalak R (2002) The health effects of indoor air pollution exposure in developing countries. Switzerland: Technical report, World Health Organization Geneva.

60. World Health Organization (2010) WHO guidelines for indoor air quality: selected pollutants. Copenhagen, Denmark: Technical report, World Health Organization, Regional Office for Europe.

Chapter 2

The Effect of Drilling Mud Properties on Shallow Lateral Resistivity Logging of Gas Hydrate Bearing Sediments

Jiaxin Sun[a], Fulong Ning[a], Nengyou Wu[b], Shi Li[c],
Ke Zhang[c], Ling Zhang[a], Guosheng Jiang[a],
and V.F. Chikhotkin[a]

[a]Faculty of Engineering, China University of Geosciences, Wuhan 430074, China

[b]Guangzhou Institute of Energy Conversion, Chinese Academy of Sciences, Guangzhou 510640, China

[c]State Key Laboratory of Enhanced Oil Recovery, Beijing 100083, China

ABSTRACT

Resistivity logging is one of the most important ways of identifying and estimating the saturation level of gas hydrates in permafrost and ocean regions. In practical drilling operations, resistivity loggings, especially shallow lateral resistivity logging in gas hydrate bearing sediments (GHBS), are likely to be affected by drilling mud invasions. Here, we use available data from site measurements to construct a two-dimensional model for hydrate reservoirs around the borehole SH2, a hole that was drilled during the first expedition in Shenhu area of South China Sea to examine and drill into gas hydrates. We then use this model to investigate the characteristics of drilling mud invasions and the effect of drilling mud properties (e.g., temperature, density, and salinity) on resistivity logging using a numerical simulation. This simulation and associated calculations indicate that shallow lateral resistivity logging is significantly affected by variations in drilling mud temperature, which leads to hydrate dissociation and the formation of secondary hydrates. Increasing drilling mud salinity accelerates hydrate dissociation, and has a greater effect on shallow lateral resistivity logging than the free gas produced during drilling and any potential mud density influence, which is generally dependent on the depth at which the drilling mud invasion occurred. This means that future drilling operations should focus on ensuring that the temperature, salinity, and density of drilling muds remain within a reasonable range in order to minimize the effect of mud invasions on resistivity logging data.

INTRODUCTION

Gas hydrates are non-stoichiometric inclusion compounds formed when hydrophobic gas molecules (usually methane and carbon dioxide) come into contact with water (host molecules) under low-temperature and high-pressure conditions (Sloan, 2003). Gas hydrates are widely distributed in areas of permafrost and in marine sediments at depths of >300 m below the seafloor. The exhaustion of traditional oil and gas resources, combined with a continuous increase in consumption, means that unconventional energy sources, such as natural gas hydrates, are considered to be the most promising future sources of energy. Klauda and Sandler (2005) state that 74,000 Gt of CH_4 is trapped in gas hydrates within marine zones, three orders of magnitude larger than current worldwide conventional natural gas reserves. Consequently, the exploration and exploitation of marine gas hydrates have become a hot topic for current and future energy research.

The main methods of exploration of marine gas hydrates are geology, geophysics, geochemistry, and core drilling based investigations. Although geophysical approaches including seismic detection (Riedel et al., 2002), well logging (Collett, 2001), and the latest marine electromagnetic technology (Schwalenberg et al., 2010) are the most widely used method of gas hydrate exploration (Song et al., 2002), core drilling is the most direct way to identify and evaluate marine gas hydrate reservoirs. A significant amount of marine gas hydrate drilling and corresponding well logging has been undertaken in a number of oceanic zones worldwide. Although a number of advances have been made, there are still three unfavorable factors (i.e., poorly characterized reservoirs, unreliable production technology, and high risks) that have hindered current gas hydrate exploration and exploitation (Ning et al., 2012). The risk factor refers to issues including drilling safety (such as wellbore instability), geological disasters, environment impacts of gas hydrate release, and formation deformation in the production processes. This indicates that ensuring safe drilling is a key issue in potential gas hydrate extraction. Previous research has indicated that overbalanced drilling, involving pressures no higher than fracture pressure, is the preferred option for marine gas hydrate drilling (Ning et al., 2008, Collett et al., 2009 and Birchwood and Noeth, 2012). This approach uses water-based drilling mud that displaces primary pore fluids (air, water) around the borehole and invades the gas hydrate bearing sediments (GHBS) as a result of hydraulic pressure gradients. It differs from the mud invasions encountered during drilling into conventional oil and gas formations, as drilling into GHBS may also be accompanied by the formation and dissociation of hydrate as a result of the frictional heat generated by the drilling tool and the relatively high temperature of the drilling mud (Ning et al., 2012 and Ning et al., 2013a; Fig. 1). The fact that dissociated GHBS-derived methane and reformed hydrates are thought to act as insulating mediums, and that GHBS salinities are also reduced by hydrate dissociation, means that drilling mud invasions have significant effects on resistivity logging undertaken during drilling, especially on shallow lateral resistivity logging, potentially inducing errors in logging interpretation and identification (Ning et al., 2012 and Ning et al., 2013a). Therefore, use of overbalanced drilling when investigating marine gas hydrate deposits means that mud invasions into GHBS and the possibly associated dissociation and reformation of gas hydrates are most commonly observed during marine hydrate exploration and exploitation drilling. Previous research undertaken during the GMGS-1 gas hydrate drilling project in the South China Sea focused on the use of TOUGH+HYDRATE software (Moridis et al., 2008) to simulate the one-dimensional invasion of drilling muds with different properties into GHBS, using gas hydrate reservoirs within the SH7 borehole as a case study. We also studied the processes involved in mud invasions and discussed the general influence of these invasions on resistivity logging and well stability (Ning et al., 2013a). Here, we focus on another borehole, SH2, establish a two-dimensional analytical model for this borehole, and further analyze the characteristics of drilling mud invasions and the effects of these invasions on well logging identification and hydrate formation assessment. The results of this study will contribute to improve the theoretical interpretation of well logging inversion data and enable more accurate correction of mud invasion-induced well logging distortions.

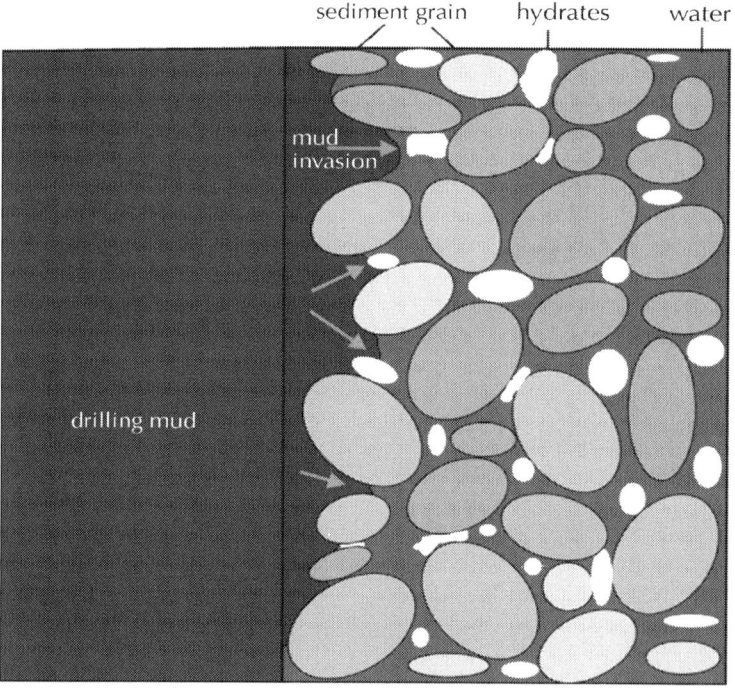

Figure 1: Schematic diagram of mud invasion into a GHBS (Ning et al... 2013b).

RESEARCH METHODS

Background

The study area is located in the southeast of the Shenhu Underwater Sandy Bench area of the central part of the north slope of the South China Sea, between the Xisha Trough and the Dongsha Archipelago. The first Chinese expedition to drill gas hydrates, GMGS-1, was undertaken in this area during April and June 2007 on behalf of the Guangzhou Marine Geological Survey (GMGS) and the Ministry of Land and Resources of the PR China (Fig. 2; Zhang et al., 2007). A total of eight sites were drilled and well logged during this project, with cores recovered at five of these sites, including three sites with recovered gas hydrate samples (SH2, SH3, and SH7; Wu et al., 2007, Wu et al., 2008 and Zhang et al., 2007). Core sample analysis indicates the presence of gas hydrates at depths of 153–229 m beneath the seafloor, with thicknesses of 10–43 m and porosities of 33–48%, in areas with water depths of 1108–1245 m. These type I methane hydrates with 26–48% saturation are disseminated throughout the sediment, and the gas produced from these hydrates was originally derived from microorganisms and consists of 96.1–99.82% methane. *In situ* measurements indicate a bottom-water temperature of 3.3–3.7 °C, with a geothermal gradient of 43–67.7 °C km^{-1}, corresponding to a sea-bottom heat flow of 74.0–78.0 mW m^{-2} (average of 76.2 mW m^{-2}).

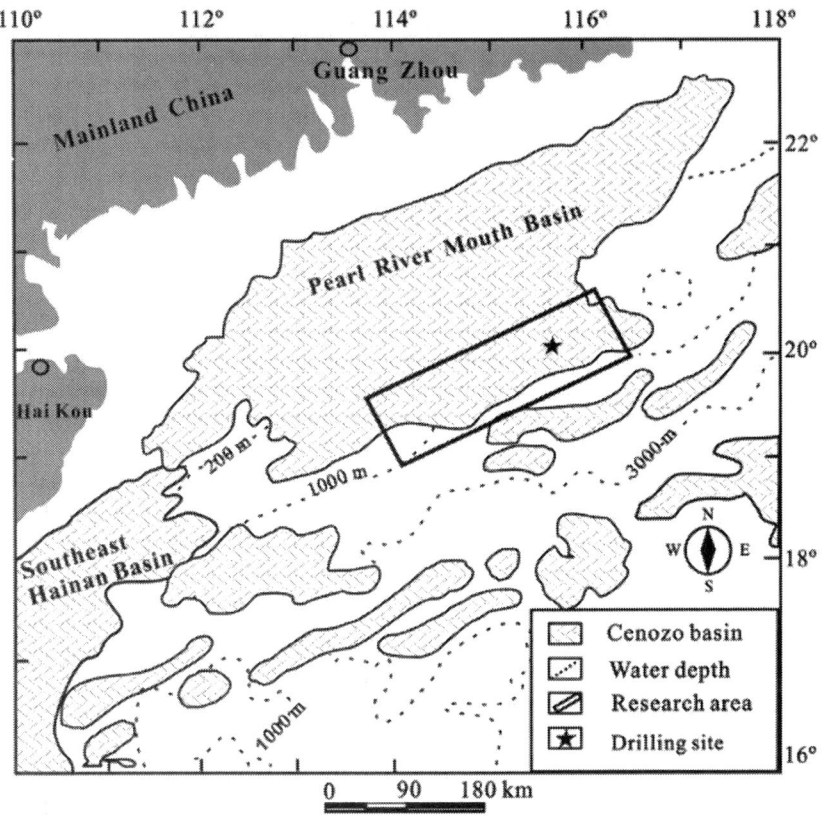

Figure 2: Location of the Shenhu GMGS-1 project within the South China Sea (Wu et al., 2009).

Riserless drilling is undertaken on a vessel during the GMGS-1 project on the vessel. The used drilling assembly at Site SH2 is shown in Table 1. The rotation rate of drill pipes is between 70-80 r min⁻¹. Seawater is used as a drilling mud with a displacement of 800–900 L min⁻¹. Seawater in the open-hole annulus is displaced by bentonite mud with a density of 1190 kg m⁻³ to improve wellbore stability when the well reaches the desired depth. The borehole is characterized by well logging and coring (Hu et al., 2009), with mud invasions generally restricted to stages of mud displacement that have only short periods of mud circulation and the preparation stage for well logging, meaning that mud cake effects can be neglected.

Table 1: The bottom hole assembly used at Site SH2 in the Shenhu area (Hu et al., 2009)

Main bottom hole assembly	Diameter (mm)	Number	Attached Equipment
Blade core bit	228.6	1	Long-sealing pup joint
Drill collar	177.8	8	Short-sealing pup joint
Drilling pipe	127	22	Landing nipple
Aluminum drilling pipe	175	100	Shut-off nipple
Drilling pipe	127	1	Floating valve nipple

Feild and/or experimental investigations of the general characteristics of mud invasion into GHBS are very difficult in present research conditions. Therefore, numerical simultion is the optimum means of understanding mud invasion into GHBS. Here we used TOUGH+HYDRATE simulation software for this study; this software was developed from TOUGH V2.0 by the Lawrence Berkeley National Laboratory, and has generally been used to simulate gas recovery from hydrate reservoirs in marine and permafrost regions (Moridis et al., 2007a, Moridis et al., 2009 and Li et al., 2010). This software can also be used with other software (e.g., FLAC 3D) to simulate wellbore stability and sediment deformation during gas production from hydrate reservoirs (Rutqvist et al., 2009), and it includes both equilibrium and kinetic models of hydrate formation and dissociation. These models cover four phases (gas, liquid, ice, and hydrate) and four mass components (water, methane gas, hydrate, and water-soluble inhibitors such as salts or alcohols) with each component existing in each phase (Moridis et al., 2008). TOUGH+HYDRATE can also simulate non-isothermal hydration reactions, phase behavior, and fluid and heat flow under conditions generally found within natural CH_4-hydrate reservoirs in complex geologic media (Zhang et al., 2009). The majority of the data used in this study were directly obtained from the SH2 site, with the rest taken from the published literature (Li et al., 2010 and Su et al., 2010). We also used the equilibrium model of hydrate formation and dissociation without allowing for chemical and mechanical coupling of the diffusion effect, and simplified our modeling by assuming that NaCl was the only thermodynamic inhibitor within the mud.

Model Construction

The simulations presented here are based on GHBS intercepted at site SH2, where the seafloor is at a water depth of 1235 m and is at a temperature of 3.7 °C. The GHBS in this area is located ~195–220 m below the seafloor (mbsf), has a thickness of 25 m, and pore water salinity (mass fraction) of 3.05%. An axisymmetric cylinder with a radius of 5 m is adopted for the model domain, which is enough to investigate the effects of drilling mud invasion acccording to the previous 1D studies (Ning et al., 2012 and Ning et al., 2013a). The borehole is located in the center of the cylinder. The drilling tool assembly mentioned before (Hu et al., 2009) is represented by a 228.6 mm diameter borehole, a 177.8 mm drilling pipe, an annular with a clearance of 25.4 mm, and bentonite mud with a density of 1190 kg m^{-3}. Hydrate saturation estimates based on pore water freshening variations with depth indicate a gradual increase in hydrate saturation at depths of 204.5–211 mbsf (shown as a shaded region in Fig. 3). In addition, the drill core obtained from this depth indicates that this area contains silty clay. As such, the GHBS at 204.5–211 mbsf, with a thickness of 6.5 m, was selected for study. The coexisting methane hydrates and water within the *in situ* sediment can be subdivided into three layers on the basis of saturation distribution from top to bottom; namely GHBS$_1$-2.5 m, GHBS$_2$-2 m, and GHBS$_3$-2 m, which have hydrate saturation values of 0.30, 0.35, and 0.40, respectively. According to Fig. 3, the three sections with successional and concentrated hydrate saturation distribution are beneficial to construct a more practical 2D model.

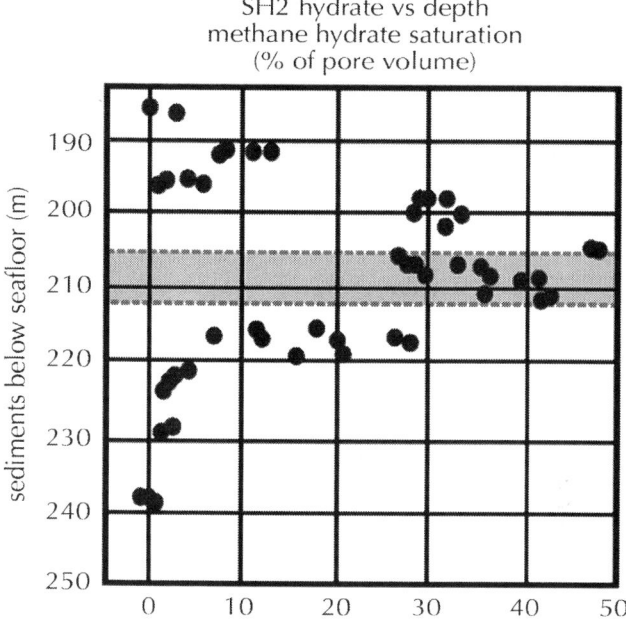

Figure 3: Hydrate saturation vs. depth at site SH2 (Nakai et al., 2007). The shadow area in purple denotes the location to be studied.

The mud in the borehole is considered as a single group and is treated as a fixed inner boundary which has stable temperature and pressure in the model (Fig. 4). The temperature of drilling mud at the depth of 204.5 mbsf is set and the distribution of this mud is assumed to be in accordance with the geothermal gradient in this area. The mud pressure is estimated from the pressure of overlying sea water and mud in the borehole, as follows:

$$P_f = P_{atm} + g\left(\rho_{sw}h + \rho_f z\right) \times 10^{-6} \tag{1}$$

where Pf is mud pressure in MPa, f is mud density in kg m^{-3}, P_{atm} is standard atmospheric pressure at 0.101325 MPa, h is water depth in m, z is the depth of sediment from the seafloor in m, g is acceleration due to gravity in m s^{-2}, and $_{sw}$ is average sea water density in kg m^{-3}; this term is a function of water depth, temperature, and salinity, and can be usually taken as 1035 kg m^{-3} (Li et al., 2010). Eq. (1)enables the calculation of the mud pressure within the borehole. The wet thermal conductivity λs and dry thermal conductivity λ_{hs} of the GHBS are taken to be 3.1 and 0.85 W m⁻¹ °C⁻¹, respectively. In addition, the density ρ of the GHBS is assumed to be 2600 kg m⁻³, and GHBS₁, GHBS₂, and GHBS₃ are assigned porosities of 42%, 49%, and 48%, respectively (Wang et al., 2010) based on density-logging curve calculations, with an intrinsic permeability of 1.0×10^{-14} m² (=10 mD; Su et al., 2010). The modeling and drilling mud parameters are given in Table 2 and Table 3, respectively.

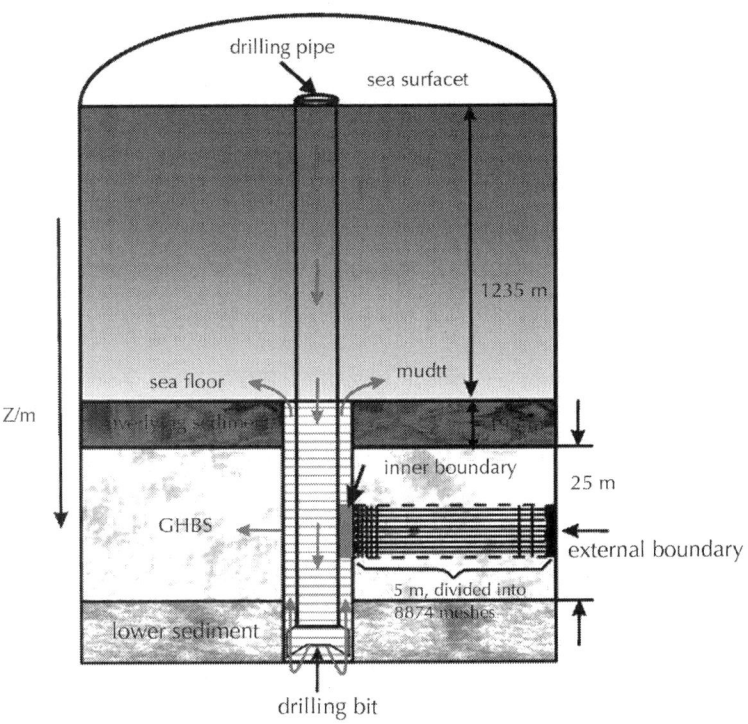

Figure 4: Schematic mud-invasion model at site SH2 of the GMGS-1 project. The drilling mud was directly discharged into sea water through the "open hole". The dashed rectangles in the GHBS represent the study area, which was divided into 8874 grid cells.

Table 2: Main properties and conditions of the drilling mud used and the hydrate deposits intercepted at Site SH2

Parameter	Value	Parameter	Value
GHBS$_1$ thickness	2.5 m	Porosity of GHBS$_2$ (φ_2)	0.49
GHBS$_2$ thickness	2.0 m	Porosity of GHBS$_3$ (φ_3)	0.48
GHBS$_3$ thickness	2.0 m	Compression coefficient a	1.00×10^{-8} Pa^{-1}
Initial top temperature of GHBS$_1$ (Ts)	14.49 °C	Grain density (ps)	2600 kg m^{-3}
Initial top pressure of GHBS$_1$ (Ps)	15.01 MPa	Geothermal gradient	46.95 K km^{-1}
Water salinity (Xw)	3.05%	Grain specific heat (Cs)	1000 J kg^{-1} °C^{-1}
Hydrate saturation in GHBS$_1$ (SH_1)	0.30	Wet thermal conductivity (λs)	3.1 W m^{-1} °C^{-1}
Pore water saturation in GHBS$_1$ (SA_1)	0.70	Dry thermal conductivity (λHs)	0.85 W m^{-1} °C^{-1}
Hydrate saturation in GHBS$_2$ (SH_2)	0.35	Thermal conductivity of hydrate (λH)	0.5 W m^{-1} °C^{-1}
Pore water saturation in GHBS$_2$ (SA_2)	0.65	Intrinsic permeability (K)	10×10^{-15} m^2 (=10 mD)
Hydrate saturation in GHBS$_3$ (SH_3)	0.40	Hydrate density (ρH)	920 kg m^{-3}
Pore water saturation in GHBS$_3$ (SA_3)	0.60	Specific heat of hydrate (CH)	2100 J kg^{-1} °C^{-1}
Porosity of GHBS$_1$ (φ_1)	0.42	Specific heat of mud (CM)	4000 J kg^{-1} °C^{-1}

Table 3: Drilling mud parameters used during simulations

Parameters of drilling mud	Value			
Invasion temperature at the top of GHBS (°C)	15*	16	17	18
Invasion density of drilling mud (kg m⁻³)	1190*	1290	1390	1490
Invasion salinity of drilling mud (%)	3.05*	6.10	12.20	18.30

Note that the values marked "*" in the table are practical drilling mud values.

The simulation uses a relative permeability model as follows (Moridis et al., 2008):

$$k_{rA} = \left(\frac{S_A - S_{irA}}{1 - S_{irA}}\right)^{n},$$

$$k_{rG} = \left(\frac{S_G - S_{irG}}{1 - S_{irA}}\right)^{nG},$$

$$k_{rH} = 0 \tag{2}$$

where *SirA* is 0.30, *SirG* is 0.05, and *n* and *nG* are 3.572, the other parameters are listed in Table 2 (the same below).

This modeling also used the following capillary pressure model (Van Genuchten, 1980):

$$P_{cap} = -P_s\left[(S^*)^{-1/\lambda} - 1\right]^{1-\lambda},$$

$$S^* = \frac{(S_A - S'_{irA})}{(S_{mxA} - S'_{irA})},$$

$$-P_{max} \leq P_{cap} \leq 0 \tag{3}$$

where   is 0.45, $_{S'irA}$ is 0.29, *SmxA* is 1.0, and P_{max} is 10^5 Pa.

The composite thermal conductivity model used in the modeling is as follows (Moridis et al., 2005):

$$\lambda_c = \lambda_{Hs} + \left(\sqrt{S_A} + \sqrt{S_H} \right) (\lambda_s - \lambda_{Hs}) + \varphi S_I \lambda_I$$

(4)

There is no ice present in the South China Sea, and, as such, SI is 0.

Domain Discretization

Fig. 4 shows the meshes employed in simulating the drilling mud invasion into GHBS. The study area is discretized into 8874 (102×87) elements in a cylindrical coordinates system (r, Z), with 8700 active elements and the rest assigned as boundary cells located in the periphery of an area at constant temperatures and pressures during the simulation. The scale of discretization varies from fine discretization (Z=0.05 m) along the Z axis in areas where hydrate saturation changes to coarser (Z=0.1 m) in other areas. This type of discretization is important and ensures accurate predictions (Moridis et al., 2007). Previous research has also indicated that phase transitions, and heat and mass transference around the borehole are rapid, meaning that very fine discretization along the r direction was used in this region, yielding the grids that include 34,800 (8700×4) coupled equations that are solved simultaneously.

Initial Conditions

The initial conditions in the reservoir follow the initialization process described by Moridis et al., (2007);Moridis and Reagan (2007, 2007b). The fact that natural gas hydrates in the Shenhu area of the South China Sea are distributed in poorly consolidated sediments near the seafloor means that pore water in the sediments can exchange with the sea-bottom water, indicating that the sediment pore water pressure is hydrostatic (Hyndman et al., 1992). That is to say, the initial hydrostatic pore water pressure can be calculated using the following empirical formula (Song et al., 2002):

$$P_{pw} = P_{atm} + \rho_{sw} g(h+z) \times 10^{-6}$$

(5)

where P_{pw} is the hydrostatic pore water pressure in MPa, and all other remaining parameters are as defined above. The water depth at site SH2 is 1235 m, so the pressure distribution of the entire system, including the pressure Ps (at $Z=-204.5$ m), can be determined. The hydrate pressure–temperature ($P–T$) equilibrium curve is then used to provide the lower limit of TS at the top of GHBS (i.e., the equilibrium T) to maintain the stability of the initial GHBS. Here, we use an initial Ts value that is slightly lower than the corresponding equilibrium temperature. This, combined with the known geothermal gradient of the GHBS listed in Table 2, means that the initial temperature at the top and bottom boundaries of the model can be determined; this temperature distribution was calculated quickly by the software for a short-duration simulation.

RESULTS AND ANALYSIS

Shallow Lateral Resistivity in Borehole SH2

The identification and assessment of hydrate reservoirs using resistivity logging assumes that gas hydrate within pores is non-conducting, in a similar function to free gas and ice. This indicates that GHBS have relatively high resistivity anomalies compared with water-saturated sediments, with these anomalies being associated with gas hydrate saturation. Hydrate saturation is usually estimated by Archie's formula (Collett, 1993 and Collett and Ladd, 2000), which is used during conventional oil/gas reservoir resistivity logging:

$$S_w = \left(\frac{aR_w}{\phi^m R_t} \right)^{1/n}, \quad S_H = 1 - S_w$$

(6)

where Rw is the resistivity of connate water, a and m are Archie constants, ϕ is porosity, n is an empirically derived saturation index, and Rt is the resistivity of the formation containing gas hydrate or other hydrocarbons. Eq. (6) can also be rewritten as follows:

$$R_t = \frac{aR_w}{\phi^m (1 - S_H)^n}$$

(7)

Eq. (7) indicates that GHBS resistivity is dependent on pore water resistivity and hydrate saturation. If a mud invasion is accompanied by hydrate dissociation, then the pore water resistivity around the borehole is affected by a change in temperature and the diluting effect of hydrate dissociation as well as the salinity of the drilling mud. In addition, hydrate dissociation will cause dynamic changes in hydrate saturation and may also produce free gas. It means that the $1-SH$ term in Eq. (7) should be replaced by $1-SH-Sg$, indicating that the apparent resistivity of the formation can be obtained using an equation based on hydrate saturation, gas saturation, and formation water resistivity during mud invasion. The results from this calculation method are then compared with the actual measured apparent resistivity of the formation to determine whether the mud invasion into GHBS has occurred, and the amount of associated hydrate dissociation.

Logging reports from GMGS-1 indicate that wire logging at site SH2 was conducted within about 20 h after drilling (Nakai et al., 2007). This indicates that the temperature, pressure, and hydrate saturation and salinity distribution in GHBS around borehole SH2 were acquired by modeling a mud invasion about 20 h. The simulated results show that no free gas produced, and the corresponding results are shown in Fig. 5.

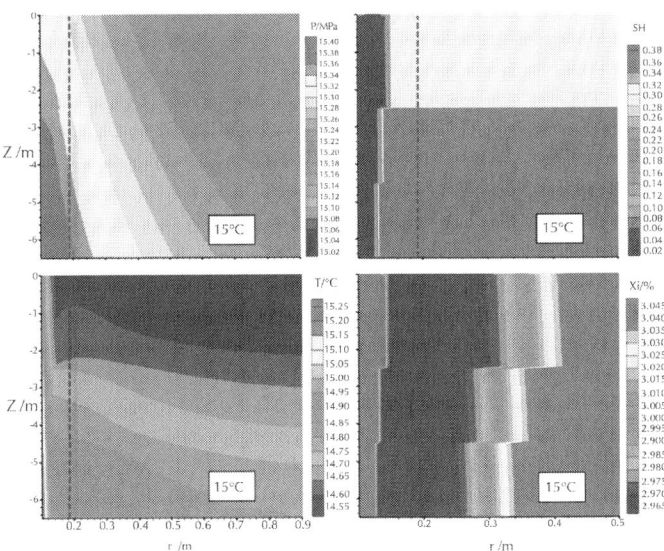

Figure 5: Distribution of physical properties of GHBS around a simulated borehole (t=20 h); the dashed lines indicate the detection depth of shallow lateral resistivity (the same below).

Fig. 5 shows that drilling mud invasion and hydrate dissociation occur in the vicinity of the borehole (within the radius of 0.5 m) within 20 h of logging. In practical logging operations in SH2, a dual-laterolog tool was used to measure both deep and shallow lateral resistivity plus spontaneous potential. The depths of investigation for deep and shallow lateral resistivity are 0.91 and 0.19 m, respectively (Nakai et al., 2007), indicating that shallow lateral well logging may be strongly distorted by drilling mud invasion, hydrate dissociation, and possible hydrate reformation, whereas deep lateral well logging is only weakly affected by these processes. The Fofonoff's Eq. (8) (Fofonoff, 1985) or the fitting Eq. (9) of NaCl solution about the resistivity with its concentration and temperature (Wang, 2013) are used to calculate the resistivity of connate water as follows:

$$\begin{cases} R = C(S,\ T,\ P)/C(35,\ 15,\ 0) \\ R = r_T(T)R_T(S,\ T)R_P(R,\ T,\ P) & -2 \le T \le 35\ ^\circ\text{C}, \\ r_T(T) = C(35,\ T,\ 0)/C(35,\ 15,\ 0) & 0 \le P \le 100\ \text{MPa}, \\ R_T(S,\ T) = C(S,\ T,\ 0)/C(35,\ T,\ 0) & 0.2\% \le X_i \le 4\%. \\ R_P(R,\ T,\ P) = C(S,\ T,\ P)/C(S,\ T,\ 0) \end{cases} \tag{8}$$

$$\begin{cases} C_w = \rho_{sw}X_i \times 10^6 \\ R_{w0} = 0.0123 + \dfrac{3627.54}{C_w^{0.955}} & X_i \ge 4\%. \\ R_w = \dfrac{45.5R_w'}{T+21.5} \end{cases} \tag{9}$$

where R is the conductivity ratio, S ($=1000 \cdot X_i$) is salinity, T is temperature in °C, P is pressure in dbar (1 dbar=0.01 MPa), C ($=1/R_w$) is conductivity in S m^{-1} and the value of conductvity $C(35, 15, 0)$ is 4.2914 S m^{-1} (Culkin and Smith, 1980). C_w is the salinity in mg L^{-1}, R_{w0} is the resistivity of NaCl solution at 24 °C and R_w is the resistivity of NaCl solution at different temperatures.

Substituting the salinities, temperatures and pressures of shallow positions obtained from the simulation into Eqs. (8) and (9) gives a value for R_w, with R_t given by substituting the corresponding hydrate and gas saturation values into Eq. (7). Values of 1.1, 2.07, and 2 are used for the Archie constants a, m, and n, respectively (Wang et al., 2010), in calculations undertaken during this study. The results of our simulation are consistent with the actual logging curve obtained shortly after drilling (Fig. 6), with any slight differences potentially due to the heterogeneity of the GHBS and neglect of clay content in the formation. In addition, some simplifications and parameters employed in this simulation can also cause these differences.For example, the heterogeneity of the GHBS is a very significant factor because the Archie constants *a* and *m* are determined by the cross plot of the density porosity and formation factors using the whole formation data. However, the study area is only a typical local layer. In addition, the saturation exponent *n* used is also a empirical value, and the mean value of formation porosities are synthetically evaluated by the density logging and the neutron porosity logging. Therefore, the parameters used in the calculation may be not fully consistent with those of study area, which may also result in the discrepancy.

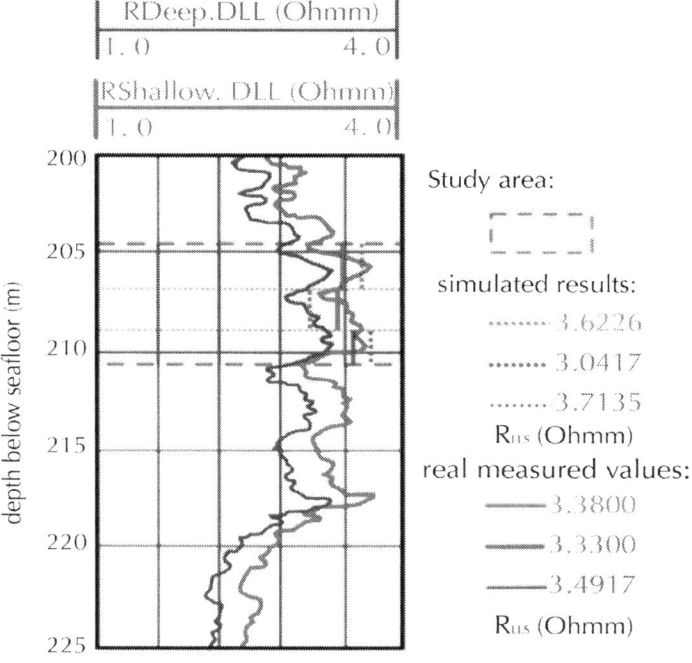

Figure 6: Well logging curves and calculated formation resistivity for the SH2 borehole; the shaded region indicates the area considered during this study (Nakai et al., 2007).

Besides, the mud salinity is equal to the initial pore water salinity of *in situ* GHBS, which implies the shallow lateral resistivity should be equal to deep one. However, resistivity logging results show that the former is higher than the later one within borehole SH2 (Fig. 6). Therefore, we think it is mud invasion coupled with hydrate dissociation and reformation resulting in the difference between shallow and deep lateral resistivities.

Previous one-dimensional studies (Ning et al., 2012, Ning et al., 2013a and Ning et al., 2013b) have shown that mud invasion and hydrate dissociation have a significant effect on the shallow lateral logging. The level of mud invasion and hydrate dissociation is associated with the properties of the drilling mud used, especially temperature, density, and salinity (Ning et al., 2012, Ning et al., 2013a and Ning et al., 2013b). Given this, we use two-dimensional simulations and the SH2 borehole as a case study to analyze the effect of drilling mud temperature, density, and salinity on mud invasions as well as shallow lateral resistivity logging.

Effect of Drilling Mud Temperature on Shallow Lateral Resistivity Logging

Fig. 7 shows the distribution of gas hydrate, free gas, and salinity in the 0.5 m around the SH2 borehole (20 h) at different drilling mud temperatures but with a fixed density (1190 kg m^{-3}) and salinity (3.05%). The simulation results shown in Fig. 7a indicate that no obvious secondary hydrate formed in the dissociation front at an invasion temperature of 15 °C at the top of GHBS (Z=−204.5 m). Increasing mud temperatures are coincident with an increase in the depth of the mud invasion, as increased amounts of hydrate dissociation cause an increase in permeability, which in turn increases the depth and extent of the mud invasion, possibly even surpassing the shallow lateral resistivity detection range (0.19 m). In addition, the formation of secondary hydrate increases hydrate saturation at the dissociation front above that of the original formation; the higher the initial hydrate saturation of a GHBS, the more secondary hydrate is produced. Secondary hydrate formation leads to an increase in shallow lateral resistivity values, but total dissociation of hydrates within the detection range will lead to a significant decrease in resistivity. Furthermore, the distribution of Sg values shown in Fig. 7b indicates that no free methane gas is produced during invasion of relatively low temperature mud, whereas hydrate dissociation is accelerated dramatically during invasion of higher temperature drilling mud, forming a significant amount of free gas at the dissociation front. This free gas is the source of the secondary hydrate discussed above. Free gas generation is similar to hydrate saturation distribution described above in that the higher the initial hydrate saturation and the hotter the mud, the more free gas is generated by hydrate dissociation. This leads to a positive correlation between increasing shallow lateral resistivity logging values within the detection range and increases in original hydrate saturation values and mud temperatures. The dissociation and reformation of hydrate may also cause changes in pore water salinity close to the borehole (Fig. 7c), with dissociation-induced local salinity reductions increasing the resistivity of the formation, a process that will increase shallow lateral resistivity logging values to a certain extent. This analysis was verified by using Eq.(7) to conduct a simple quantitative calculation In the calculation, we just use the simulation results at the center position of each layer because they vary very small in the whole layer. The results are given in Table 5.

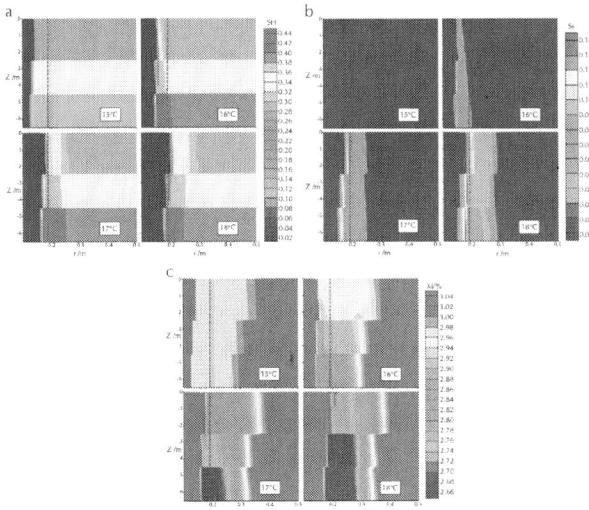

Figure 7: Changes in physical properties of GHBS around a simulated bore-hole at different mud temperatures (t=20 h).(a) Distribution of hydrate satu-

ration around a simulated borehole. (b) Distribution of free gas saturation around a simulated borehole. (c) Distribution of NaCl concentrations around a simulated borehole.

Table 4: Parameters used during resistivity calculations

Depth (m)	Mud temperature (°C)	Pressure (MPa)	Temperature (°C)	Salinity (%)	Initial hydrate saturation	Hydrate saturation (20 h)	Gas saturation
205.75	15	15.3113	14.6095	2.9662	0.3000	0.2991	0
208		15.3355	14.7608	2.9649	0.3500	0.3489	0
210		15.3565	14.8789	2.9639	0.4000	0.3988	0
205.75	16	15.3053	15.0612	2.9534	0.3000	0.2996	0
208		15.3219	15.1115	2.9119	0.3500	0.3491	0
210		15.3435	15.1602	2.8535	0.4000	0.4101	0.0747
205.75	17	15.3234	15.1615	2.8289	0.3000	0.3346	0.0877
208		15.3417	15.2125	2.7428	0.3500	0.3754	0.0873
210		15.3611	15.2695	2.6454	0.4000	0.4224	0.0839
205.75	18	15.3286	15.4267	3.0371	0.3000	0	0.0928
208		15.3554	15.2673	2.6438	0.3500	0.3842	0.0907
210		15.3737	15.3605	2.4691	0.4000	0.4318	0.0866

Table 5: Results of formation resistivity calculations undertaken at depths of 205.75, 208, and 210 m during this study

Mud temperature (°C)	Resistivity (Ω m)	205.75 m	208 m	210 m
15	Rt	3.6226	3.0417	3.7135
16		3.6043	3.0676	5.1986
17		5.4953	4.7409	6.0462
18		2.0774	5.1242	6.7485

These results indicate that an increase in mud temperature causes an initial increase then decrease in the shallow lateral resistivity of a formation with low initial hydrate saturation. In comparison, a formation with high initial hydrate saturation undergoes a continuous increase in resistivity. This continuous increase is the result of significant amounts of secondary hydrate formation and free gas production within the detection range, both of which are caused by accelerated hydrate dissociation associated with the increased mud temperature. In addition, free water released during hydrate dissociation has a diluting effect on pore water salinity, especially within formations with high levels of hydrate saturation. However, very high mud temperatures cause complete dissociation of hydrates within the detection range at the top of GHBS, an area with lower initial hydrate saturation (Fig. 5a), causing a significant reduction in resistivity.

Effect of Drilling Mud Density on Shallow Lateral Resistivity Logging

Overbalancing is usually used to ensure safe drilling in marine gas hydrate formations. Increasing the pressure in the borehole causes an increase in hydraulic pressure gradient between the borehole and formation, which finally affects drilling mud invasion. Fig. 8 shows the main physical parameters (hydrate saturation and salinity; the simulation undertaken during this study indicates a lack of free gas production) of formations with identical temperature (15 °C) and salinity (3.05%) conditions, but with differing mud densities. These data indicate that an increase in mud density does not result in the formation of secondary hydrate around the borehole (Fig. 8a). This lack of secondary hydrate is independent of the original hydrate saturation of the simulated formation. However, the range of hydrate dissociation increases with decreasing formation hydrate saturation, primarily as a reduction in hydrate saturation increases permeability and enhances both the depth of mud invasion and the extent of dissociation, as does an increase in drilling mud density. Such an increase in density also increases the hydraulic pressure gradient. The borehole pressures are still slightly lower than the equilibrium pressure at this mud temperature even though the mud density increases in this simulation. Logging values may also decrease significantly once the mud invasion depth goes beyond the detection range of shallow lateral resistivity logging. This is shown in Fig. 8a, resistivity logging results are controlled by whether the area of complete hydrate dissociation is within or outside of the logging detection range when drilling mud salinity is equal to the initial salinity of pore water. Fig. 8b shows salinity values within 0.9 m of a borehole and indicates the effect of varying mud densities, where an increase in density is associated with an increase in the area of pore water salinity reduction; to be specific, lower hydrate saturation values yield larger areas of decreased salinity. However, pore water salinities within the detection range of shallow lateral resistivity logging are relatively uniform, indicating that any change in resistivity caused by changes in pore water salinity is rather small. The simulation undertaken during this study indicates that the effect of drilling mud density variations is dependent on whether hydrate dissociation occurs within or beyond the resistivity logging detection range, primarily as no clear secondary hydrate or free gas was generated during the simulation and formation pore water resistivity within the logging detection range is relatively uniform. We also undertook a calculation based on Eq. (7), with specific calculation parameters; the parameters and results are shown inTable 6 and Table 7, respectively.

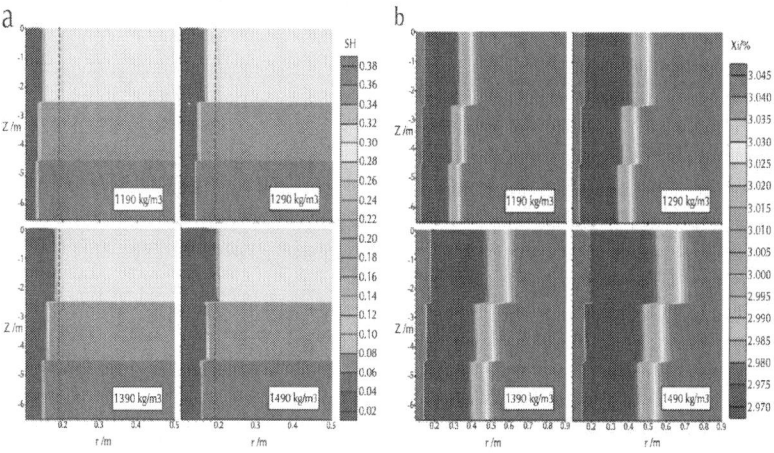

Figure 8: Distribution of physical properties of GHBS around a simulated borehole with different mud densities (t=20 h).(a) Distribution of hydrate saturation around a simulated borehole. (b) Distribution of NaCl concentrations around a simulated borehole.

Table 6: Parameters used during resistivity calculations

Depth (m)	Mud density (kg m⁻³)	Pressure (MPa)	Temperature (°C)	Salinity (%)	Initial hydrate saturation	Hydrate saturation (20 h)	Gas saturation

205.75	1190	15.3113	14.6095	2.9662	0.3000	0.2991	0
208		15.3355	14.7608	2.9649	0.3500	0.3489	0
210		15.3565	14.8789	2.9639	0.4000	0.3988	0
205.75	1290	15.5071	14.4570	2.9675	0.3000	0.2983	0
208		15.5296	14.6604	2.9658	0.3500	0.3482	0
210		15.5499	14.8357	2.9643	0.4000	0.3979	0
205.75	1390	15.7087	14.3320	2.9685	0.3000	0.2971	0
208		15.7288	14.5737	2.9665	0.3500	0.3471	0
210		15.7490	14.7235	2.9652	0.4000	0.3970	0
205.75	1490	15.9087	14.2210	3.0410	0.3000	0	0
208		15.9318	14.4822	2.9673	0.3500	0.3456	0
210		15.9506	14.6319	2.9660	0.4000	0.3959	0

Table 7: Results of formation resistivity calculations undertaken at depths of 205.75, 208, and 210 m during this study

Mud density (kg m^{-3})	Resistivity (Ω m)	205.75 m	208 m	210 m
1190	Rt	3.6226	3.0417	3.7135
1290		3.6224	3.0409	3.7055
1390		3.6199	3.0344	3.7017
1490		1.7544	3.0261	3.6941

These calculations are consistent with the simulation described above, with the resistivity of the formation decreasingly only slightly with an increase in drilling mud density, and with significant reductions in formation resistivity only occurring with significant mud invasion depths that are larger than the logging detection depths. This is mainly due to the increased mud density, which enhances the pressure difference between the borehole and the formation, thereby accelerating permeation rates and also promoting hydrate dissociation to some extent. In general, borehole pressures that are lower than the equilibrium pressure yield a situation where only large hydraulic pressure gradients and low formation hydrate saturation levels can generate large permeation depths that significantly affect logging results. This indicates that relatively small fluctuations in mud density will not significantly influence resistivity logging.

Effect of Drilling Mud Salinity on Shallow Lateral Resistivity Logging

Deep-water oil or gas drilling usually uses high concentrations of inorganic salt as a thermodynamic inhibitor to enhance drilling performance, and to prevent gas hydrate formation and aggregation within the borehole. Fig. 9a shows that mud invasion depths continue to increase under fixed mud density (1190 kg m^{-3}) and temperature (15 °C) conditions, primarily as the range of hydrate dissociation expands as a result of a decrease in initial formation hydrate saturation and an increase in drilling mud salinity. Resistivity values decrease significantly when the depth of hydrate dissociation exceeds the detection range of shallow lateral resistivity logging, in a similar fashion to the effect of high mud density described above. In addition, increased mud salinity causes a coincident increase in the amount of secondary hydrate formation at the dissociation front, a feature that is clearer in formations with higher initial hydrate saturation levels. Resistivity values will also increase if the region of increased hydrate saturation associated with the formation of secondary hydrate lies within the 0.19 m range of logging, although if the region of increased saturation lies beyond the logging detection range then the resistivity values obtained during shallow logging will be very low regardless of the high hydrate saturation levels within the formation.

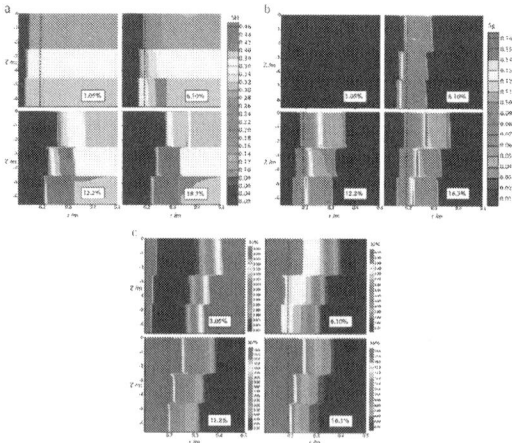

Figure 9: Distribution of physical properties of GHBS around a simulated borehole with different mud salinities (NaCl, t=20 h). (a) Distribution of hy-

drate saturation around a simulated borehole. (b) Distribution of free gas saturation around a simulated borehole. (c) Distribution of NaCl concentrations around a simulated borehole.

Fig. 9b and c indicates that higher salinity levels are associated with higher amounts of dissociation-related free gas production, a process that also changes the salinity distribution near the borehole. The presence of dissociation-related free gas within the logging detection range will increase resistivity values, but, as suggested by the previous analysis, the opposite is not true. Differing mud salinities may also cause an increase in the formation salinity, causing an associated reduction in resistivity. This indicates that drilling mud salinity variations have complex effects on shallow lateral resistivity logging, primarily as the free gas produced by dissociation will increase formation resistivity, opposing the decrease in resistivity associated with increased mud salinity. The final resistivity will depend on the relative magnitude of these two factors. To examine these results further, we used Eq. (7) to calculate resistivity values using the parameters shown in Table 8; the results of these calculations are given in Table 9.

Table 8: Parameters used during resistivity calculations

Depth (m)	Mud salinity (%)	Pressure (MPa)	Temperature (°C)	Salinity (%)	Initial hydrate saturation	Hydrate saturation (20 h)	Gas saturation
205.75	3.05	15.3113	14.6095	2.9662	0.3000	0.2991	0
208		15.3355	14.7608	2.9649	0.3500	0.3489	0
210		15.3565	14.8789	2.9639	0.4000	0.3988	0
205.75	6.10	15.3259	13.7350	5.7215	0.3000	0.1929	0.1039
208		15.3476	13.8264	5.5634	0.3500	0.3088	0.0937
210		15.3688	13.9592	5.3219	0.4000	0.3672	0.0903
205.75	12.2	15.3212	13.1608	12.1561	0.3000	0	0.0799
208		15.3522	13.1431	12.1340	0.3500	0	0.0911
210		15.3784	13.2840	11.6310	0.4000	0	0.1019
205.75	18.3	15.3075	13.1016	18.2375	0.3000	0	0.0757
208		15.3403	12.9889	18.1678	0.3500	0	0.0841
210		15.3685	13.1721	17.3821	0.4000	0	0.0926

Table 9: Results of formation resistivity calculations undertaken at depths of 205.75, 208, and 210 m during this study

Mud salinity (%)	Resistivity (Ω m)	205.75 m	208 m	210 m
3.05	*Rt*	3.6226	3.0417	3.7135
6.10		1.9437	2.0007	2.6247
12.2		0.6059	0.4522	0.4985
18.3		0.4358	0.3246	0.3552

These calculations indicate that higher drilling mud salinities are associated with larger decreases in formation resistivity values, with formations with low hydrate saturation values having low lateral resistivity logging values that are significantly affected by high salinity drilling muds, whereas high hydrate saturation formations are only affected by dramatic increases in drilling mud salinities, primarily as NaCl is a thermodynamic inhibitor. The addition of high concentrations of salt to drilling muds will promote the dissociation of hydrate; the higher the salinity of the mud, the faster the rate of hydrate dissociation. In addition, the depth of invasion will increase with any lowering of formation hydrate saturation levels. These results also indicate that any increase in pore water salinity has a more significant effect on shallow lateral resistivity values than any free gas produced by hydrate dissociation.

CONCLUSIONS AND SUGGESTIONS

This study used drilling and logging data from drilling of a marine gas hydrate deposit at site SH2 in the Shenhu area of the South China Sea by the China Geological Survey in 2007 to construct a two-dimensional model of drilling mud invasion into a gas hydrate formation. This model was combined with TOUGH+HYDRATE software to investigate the effects of drilling mud invasions and changes in mud properties on shallow lateral resistivity logging. The numerical simulations undertaken during this study yield the following results.

- Increases in mud temperatures are associated with an initial increase and then a decrease in the shallow lateral resistivity of formations with low hydrate saturation, whereas formations with higher initial hydrate saturation undergo a continuous increase in resistivity. It is likely that further increases in mud temperature will cause the complete dissociation of hydrate within the detection range of shallow lateral resistivity logging, causing a significant reduction in resistivity. This indicates that increasing but relatively low mud temperatures will be associated with an increase in shallow lateral resistivity, especially in formations with high levels of hydrate saturation. However, highly increasing mud temperatures will cause complete dissociation of hydrate within the detection range, thereby significantly reducing resistivity.

- Modeling of a scenario with a fixed temperature (15 °C) and salinity (3.05%) indicates that an increase in mud density results in only a slight reduction in formation resistivity values. However, relatively high hydraulic pressure gradients combined with relatively low levels of formation hydrate saturation can cause an increased depth of mud invasion, becoming deeper than the logging detection range and therefore causing a significant reduction in formation resistivity. This indicates that small mud density variations will generally only cause few variations in resistivity values, and will not have a significant effect on the final logging data.

- The higher the drilling mud salinity, the larger the decrease in formation resistivity when mud invasion is coupled with hydrate dissociation and secondary hydrate formation. In addition, shallow lateral resistivity logging of formations with low levels of hydrate saturation is more likely to be significantly affected by high salinity drilling mud invasions. Calculations undertaken during this study indicate that increases in formation pore water salinity have a greater effect on shallow lateral resistivity values than the presence of free gas generated by hydrate dissociation. This implies that for hydrate formation logging and borehole stability the kinetic inhibitors should be added to drilling mud rather than thermodynamic inhibitors.

The simulation results indicate that shallow lateral resistivity logging is significantly affected by the mud invasion, especially the temperature and salinity of drilling mud, but is only weakly affected by the density of drilling mud. This implies that reducing mud temperature, controlling mud density and low salinity, and the addition of temporarily blocked filtrate reducers to decrease invasion depth may be beneficial to gas hydrate drilling and reservoir evaluation, and can also have positive effects on borehole stability, improve the results of well logging, and reduce reservoir damage associated with drilling mud invasion.

ACKNOWLEDGEMENTS

The authors would like to think Dr. George Moridis and Keni Zhang for valuable suggestions that improved our modeling and theoretical analysis. We also thank two anonymous reviewers for their vaulable comments and suggestions. This work was supported by the National Natural Science Foundation of China (nos. 51274177, 40974071), Enhanced Oil Recovery State Key Laboratory Project (2011A-1002), a Key Project of the Natural Science Foundation of Hubei Province (2012FFA047) and the Fundamental Research Funds for the Central Universities (nos. CUGL100410 and 120112).

REFERENCES

1.   Birchwood, R., Noeth, S., 2012. Horizontal stress contrast in the shallow marine sediments of the Gulf of Mexico sites Walker Ridge 313 and Atwater Valley 13 and 14—geological observations, effects on wellbore stability, and implications for drilling. Mar. Pet. Geol. 34 (1), 186–208.

2.   Culkin, F., Smith, N.,D., 1980. Determination of the concentration of potassium chloride solution having the same electrical conductivity, at 15 1C and infinite frequency, as standard seawater of salinity 35.0000 ‰ (chloride 19.37394 ‰). IEEE J. Oceanic Eng. 5 (1), 22–23.

3.   Collett, T.S., Boswell, R., Mrozewski, S., Guerin, G., Cook, A., Frye, M., Shedd, W., McConnell, D., 2009. Gulf of Mexico gas hydrate joint industry project leg II—Operational summary. In: Proceedings of the Drilling and Scientific Results of the 2009 Gulf of Mexico Gas Hydrate Joint Industry Project Leg II. Available at ⟨http://www.netl.doe.gov/technologies/oil-gas/publications/ Hydrates/ 2009Reports/OpSum.pdf⟩.

4.   Collett, T.S., 1993. Natural gas hydrates of the Prudhoe Bay and Kuparuk River area, North Slope, Alaska. AAPG Bull. 77 (5), 793–812.

5.   Collett, T.S., 2001. A review of well-log analysis techniques used to assess gas- hydrate-bearing reservoirs. Nat. Gas Hydrates 124, 189–210.

6.   Collett, T.S., Ladd J., 2000. Detection of gas hydrate with downhole logs and assessment of gas hydrate concentrations (saturations) and gas volumes on the blake ridge with electrical resisitivity log data. In: Proceedings of the Ocean Drilling Program. Scientific Results, Ocean Drilling Program. pp. 179-191.

7.   Fofonoff, N.P., 1985. Physical properties of seawater: a new salinity scale and equation of state for seawater. J. Geophys. Res. 90 (C2), 3332–3342.

8.   Hyndman, R., Foucher, J., Yamano, M., et al., 1992. Deep sea bottom-simulatingreflectors: calibration of the base of the hydrate stability field as used for heat flow estimates. Earth Planet. Sci. Lett. 109 (3), 289–301.

9.   Hu, H.L., Tang, H.X., Luo, J.F., et al., 2009. Deepwater gas hydrates drilling and coring techniques. Oil Drill. Prod. Technol. 31 (001), 27–30. http://dx.doi.org/10.3969/j. issn.1000-7393.2009.01.008.

10. Klauda, J.B., Sandler, S.I., 2005. Global distribution of methane hydrate in ocean sediment. Energy Fuels 19 (2), 459–470.

11. Li, G., Moridis, G.J., Zhang, K.N., et al., 2010. Evaluation of gas production potential from marine gas hydrate deposits in Shenhu area of South China Sea. Energy Fuels 24 (11), 6018–6033.

12. Moridis, G.J., Kowalsky, M.T., Pruess, K., et al., 2007a. Depressurization-induced gas production from class-1 hydrate deposits. SPE Reservoir Eval. Eng. 10 (5), 458–481.

13. Moridis, G.J., Reagan, M.T., 2007b. Strategies for gas production from oceanic class

14. 3 hydrate accumulations. In: Offshore Technology Conference, 30 April–3 May

15. 2007, Houston, Texas.

16. Moridis, G.J., Reagan, M.T., Kim, S.J., et al., 2009. Evaluation of the gas production potential of marine hydrate deposits in the Ulleung Basin of the Korean East Sea. SPE J. 14 (4), 759–781.

17. Moridis, G.J., Kowalsky M.B., et al., 2008. TOUGHþHYDRATE v1.0 User's Manual: A Code for the Simulation of System Behavior in Hydrate-bearing Geologic Media, Lawrence Berkeley National Laboratory, Berkeley, CA. ⟨http://esd.lbl.gov/files/ research/projects/tough/documentation/TplusH_Manual_v1.pdf⟩.

18. Moridis, G.J., Reagan, M.T., 2007a. Gas production from oceanic class 2 hydrate accumulations. In: Offshore Technology Conference, 30 April–3 May 2007, Houston, Texas.

19. Moridis, G.J., Seol Y., Kneafsey T.J. et al., 2005. Studies of reaction kinetics of methane hydrate dissociation in porous media. In: Fifth International Conference on Gas Hydrates, Trondheim, Norway.

20. Nakai, T.C., Tjok, K.M., Humphrey, G., 2007. Deepwater Gas Hydrate Investigation Shenhu Survey Area South South China Sea, Offshore China. Factual Field Report. Report no. 0201-6130. Guangzhou Marine Geological Survey, Guangzhou, PR China.

21. Ning, F.L., Jing, G.S., Zhang, L., et al., 2008. Analysis of key factors affecting wellbore stability in gas hydrate formations. Pet. Drill. Tech. 36 (3), 59–61. http://dx.doi. org/10.3969/j.issn.1001-0890.2008.03.014.

22. Ning, F.L., Zhang, K.N., Wu, N.Y., et al., 2012. Invasion of water-based drilling mud into oceanic gas-hydrate-bearing sediment: one-dimensional numerical simulations. Chin. J. Geophys. 56 (1), 204–218. http://dx.doi.org/10.6038/cjg20130121.

23. Ning, F.L., Wu, N.Y., Yu, Y.B., et al., 2013b. Invasion of drilling mud into gas-hydratebearing sediments. Part II: Effects of geophysical properties of sediments. Geophys. J. Int. 193 (3), 1385–1398.

24. Ning, F.L., Zhang, K.N., Wu, N.Y., et al., 2013a. Invasion of drilling mud into gashydrate-bearing sediments. Part I: Effect of drilling mud properties. Geophys. J. Int. 193 (3), 1370–1384.

25. Riedel, M., Spence, G., Chapman, N., et al., 2002. Seismic investigations of a vent field associated with gas hydrates, offshore Vancouver Island. J. Geophys. Res. 107 (B9), 2200.

26. Rutqvist, J., Moridis, G.J., Grover, T., et al., 2009. Geomechanical response of permafrost-associated hydrate deposits to depressurization-induced gas production. J. Pet. Sci. Eng. 67 (1-2), 1–12.

27. Sloan, E.D., 2003. Clathrate hydrate measurements: microscopic, mesoscopic, and macroscopic. J. Chem. Thermodyn. 35 (1), 41–53.

28. Song, H.B., Jiang, W.W., Zhang, W., et al., 2002. Progress on marine geophysical studies of gas hydrates. Prog. Geophys. 17 (2), 224–229. http://dx.doi.org/ 10.3969/j.issn.1004-2903.2002.02.006.

29. Schwalenberg, K., Haeckel, M., Poort, J., et al., 2010. Evaluation of gas hydrate deposits in an active seep area using marine controlled source electromagnetics: results from Opouawe Bank, Hikurangi Margin, New Zealand. Mar. Geol. 272 (1), 79–88.

30. Su, Z., Moridis, G.J., Zhang, K.N. et al., 2010. Numerical investigation of gas production strategy for the hydrate deposits in the Shenhu area. In: Offshore Technology Conference, Houston, TX.

31. Van Genuchten, M.T., 1980. A closed-form equation for predicting the hydraulic conductivity of unsaturated soils. Soil Sci. Soc. Am. J. 44 (5), 892–898.

32. Wang, J.C., 2013. Study on Methods of Reservoir Evaluation and Remaining Oil Prediction based on Static and Dynamic Logging in Water Flooded Injected by Fresh Water and Sewage. Yangtze University.

33. Wang, X.J., Wu, S.G., Liu, X.W., et al., 2010. Estimation of gas hydrate saturation based on resistivity logging and analysis of estimation error. Geoscience 24 (5), 993–999. http://dx.doi.org/10.3969/j.issn.1000-8527.2010.05.022.

34. Wu, N.Y., Zhang, H.Q., Yang, S.X. et al., 2008. Preliminary discussion on gas hydrate reservoir system of Shenhu Area, North Slope of South China Sea. In: Proceedings of the Sixth International Conference on Gas Hydrates (ICGH 2008), Vancouver, BC, Canada.

35. Wu, N.Y., Zhang, H.Q., Su, X., et al., 2007. High concentrations of hydrate in disseminated forms found in very fine-grained sediments of Shenhu area, South China Sea. Terra Nostra (1-2), 236–237.

36. Wu, N.Y., Yang, S.X., Wang, H.B., et al., 2009. Gas-bearing fluid influx sub-system for gas hydrate geological system in Shenhu

Area, Northern South China Sea. Chin. J. Geophys. 52 (6), 1641–1650. http://dx.doi.org/10.3969/j.issn.0001-5733.2009.06.027.

37.   Zhang, H.Q., Yang, S.X., Wu, N.Y. et al., 2007. Successful and surprising results for China's first gas hydrate drilling expedition. In: Fire in the Ice: Methane Hydrate Newsletter, Fall 2007, pp. 6–9.

38.   Zhang, K.N., Moridis, G.J., Wu, Y.S. et al., 2009. A domain decomposition approach for large-scale simulations of flow processes in hydrate-bearing geologic media. In: Proceedings of the Sixth International Conference on Gas Hydrates (ICGH 2008), Vancouver, British Columbia, Canada.

Chapter 3

Physical Properties of Sediments from the Ulleung Basin, East Sea: Results from Second Ulleung Basin Gas Hydrate Drilling Expedition, East Sea (Korea)

J.Y. Lee [a], G.-Y. Kim[a], N.K. Kang[a], B.-Y. Yi[a], J.W. Jung[a], J.-H. Im[b], B.-K. Son[a], J.-J. Bahk[a], J.-H. Chun[a], B.-J. Ryu[a], and D.S. Kim[c]

[a]Petroleum and Marine Research Division, Korea Institute of Geoscience and Mineral Resources, Daejeon 305-350, Republic of Korea

[b]Department of Earth System Sciences, Yonsei University, Seoul 120-749, Republic of Korea

[c]Korea Gas Corporation, Seongnam, Kyeonggi-do 463-754, Republic of Korea

ABSTRACT

Physical properties, including basic index properties, undrained shear strength, and thermal conductivities were measured on conventional cores retrieved from the Ulleung Basin, during the second Ulleung Basin Gas Hydrate expedition (UBGH2). The UBGH2 logged 13 sites and cored 10 sites to locate potential sites for an offshore test production. A total of 211 conventional cores were recovered. Undrained shear strength, thermal conductivity, and index properties, including porosity and grain density, were measured onboard. Mineral composition analyses and grain size analyses were completed onshore after the expedition.

The averaged porosity at each site ranges from 65% to 71%, the averaged grain density at each site ranges from 2.57 to 2.66 g/cm^3, and the mean grain size mostly lies between 4 μ m and 25 μ m, with sparsely scattered coarse grained intervals. The relatively high porosity and low grain density are due to the large portion of diatomaceous sediments. The thermal conductivities average around 0.8 W/mK, and the low porosity and the abundance of clay mineral and OAPL-A may have caused the relatively low thermal conductivity in the Ulleung Basin.

The geomechanical analyses revealed a few relevant findings important for sediment physical behaviors during gas hydrate production. The particle migration in coarse grained layers during production necessitates proper measure for sand productions while fine grained layers are mostly self-filtering. The vertical deformation estimated from the compression index suggested that the subsidence induced by the pore pressure change from the depressurization is much higher than those induced by the lost of gas hydrate particles from dissociation. Massive vertical deformation in fine-grained layers has been predicted to occur due to the pore pressure change from depressurization induced gas hydrate production. The drilling mud weight induced pressure window is mostly determined by the water depth in deep sea drilling, but caution should be taken in determination of fracture pressures in hydrate-bearing sediments, because gas hydrate saturations alter Poisson›s ratio.

INTRODUCTION

Physical properties provide the basic understandings of the gas hydrate (GH) occurrences and associated production technologies. Establishing safe and efficient production strategies of gases from hydrate deposits requires the assessment of production rates, well bore stability, and flow assurances. Physical properties such as fluid flow characteristics, strength and deformation characteristics, and thermal conduction characteristics are elements of these assessments (Lee et al., 2011b). Reservoir fluid flow characteristics govern production rate, with the diffusion of depressurization front and the dissipation of excess pore water being important factors (Liu and Flemings, 2007, Moridis et al., 2007 and Nimblett and Ruppel, 2003). Strength and deformation characteristics govern well bore stability and integrity of formation during production (Lee et al., 2010c, Masui et al., 2008 and Rutqvist and Moridis, 2007). Thermal conduction characteristics govern gas hydrate dissociation rate, heat front diffusion in the thermal stimulation method, and the change of gas hydrate stability zone by the any external change in pressure and/or temperature (Ruppel, 2000, Waite et al., 2002 and Waite et al., 2007).

The physical properties can be obtained from both geophysical logs and core analyses. Geophysical logging gives continuous measurements on in-situ sediments but the interpretations are often dependent on the borehole conditions. Consequently, core analyses data that are not affected by the borehole conditions, are often used to validate logging data (Winters et al., 2011). However, core analyses have also disadvantage over geophysical logging, since cores experiences stress release upon recovery. Core analyses spans from basic index property measurements to sophisticated geomechanical laboratory experiments. Index properties such as porosity, grain density, and grain size are basic criteria to characterize the physical behavior of sediments during gas hydrate production. Consequently, the index properties are very useful tools for estimating geomechanical behaviors in the absence of an extensive set of the sophisticated geomechanical laboratory experiments (Lee et al., 2013).

Gas hydrates were first found in shallow gravity cores as nodular forms at a vent site in the center of the Ulleung Basin during a gas hydrate expedition using the RV Tamhae II of KIGAM (Korea Institute of Geoscience and Mineral Resources) (Kim et al., 2011). After the discovery, the Ulleung Basin Gas Hydrate Expedition 1 (UBGH1) logged 5 sites and cored 3 sites in 2007 and identified gas hydrate occurrences as grain-displacing veins in mud layers and pore-filling forms in thin-bedded turbidite sand layers (Bahk et al., 2011). The UBGH2 logged 13 sites and cored 10 sites to locate potential sites for an offshore production test. A total of 211 conventional cores were recovered during the UBGH2. Undrained shear strength, thermal conductivity, and index properties, including porosity and grain density, were measured onboard. Mineral composition analyses and grain size analyses were completed onshore after the expedition.

STUDY AREA

The Ulleung Basin is a back-arc basin in the East Sea, a marginal sea between the Eurasian continent and the Japan Arc, and is bounded by the Korean peninsula to the west, the Yamato Ridge and the Oki Bank to the East, and the Korea Plateau to the North (Fig. 1). The Ulleung Basin has been formed from Late Oligocene to Early Miocene by the crustal extension during the drift of Japan Arc away from the Asian mainland (Jolivet et al., 1995 and Chough and Barg, 1987).

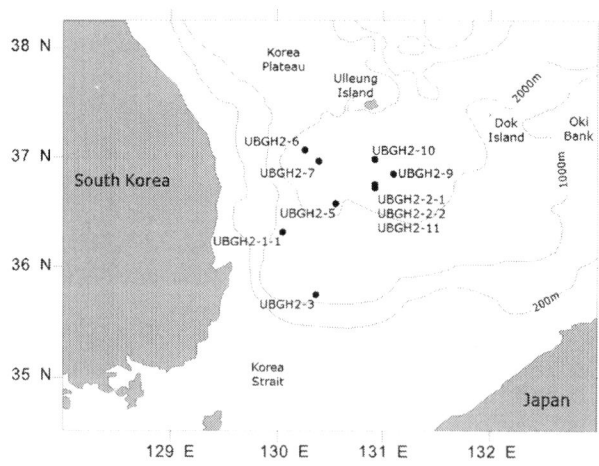

Figure 1: Locations of sites with core recovery during the 2010 UBGH2 drilling expedition in the Ulleung Basin, East Sea, Korea.

The Ulleung Basin is divided into four sequence units according to seismic studies (Chough et al., 2000). The uppermost unit is composed of a Late Pliocene–Quaternary turbidites and hemipelagic muds, and mass-flow deposits. The second unit is a Late Miocene–Early Pliocene marine shale unit, interbedded with sandstone and siltstone beds. The third unit is a Middle Miocene of marine shale unit, graded into a unit with sandstone/shale, volcaniclastics and turbidite sequence to the north east. The lowest unit is a Late Oligocene–Early Miocene unit of volcanic flows and sills, intercalated with sedimentary layers.

The UBGH2 expedition cored ten sites (Fig. 1; Ryu et al., 2012). Site UBGH2-1_1 is located on the western margin of the Ulleung Basin. Site UBGH2-2_1 is located within a seismic chimney structure in the center of the basin and is located about 1 km from Site UBGH2-2_2, a non-chimney site. Site UBGH2-3 is also located at a seismic chimney structure in the shallowest and southern-most region of the study area. Site UBGH2-5 is located on a topographic high in the north-central part of the basin. Site UBGH2-6 is the northern-most site established during UBGH2, and Site UBGH2-7, a seismic chimney site, is located near Site UBGH2-6. Sites UBGH2-9 and 2-10 are located in the north-eastern part of the basin. Site UBGH2-11 is located near Site UBGH2-2_2 and was drilled in a seismic chimney structure.

METHODS

The physical properties were measured in cores recovered with conventional wireline coring systems. The coring systems used during the UBGH2 expeditions are FHPC (Fugro Hydraulic Piston Corer), FC (Fugro Corer), and FRC (Fugro Rotary Corer). Pressurized coring systems were FPC (Fugro Pressure Corer) and FRPC (Fugro Rotary Pressure Corer). All of these systems used clear CAB liners. A total of 211 conventional cores and 29 pressure cores were retrieved from 18 holes at 10 sites. A total of 160 FHPC, 51 FC, 7 FRPC, and 22 FPC were recovered.

In-situ temperature: In-situ temperature tests were performed using the WISON EP system at 21 locations in 11 holes at 10 sites. After the temperature probe system reached a target penetration depth, the test was allowed to run for a predetermined time based on previous tests. Data were recorded and displayed in real time at the operator's workstation. At the end of the test, the tool was pulled out of the sediment formation and retrieved to deck. Detailed operation is described in the UBGH2 initial report (Ryu et al., 2012).

Moisture and density: Moisture and density analysis (MAD) was used to measure wet mass, dry mass, and dry volume to determine moisture content, grain density, bulk density, porosity and void ratio, as described in the Ocean Drilling Program (ODP) Technical Note 26 (Bium, 1997). One or two sample plugs of ~10 cm³per section were taken at or near the location of other physical property measurements.

Grain size: Grain-size analysis was conducted onshore using a laser-scattering particle analyzer (Microtrac S3500).

Mineral compositions: Mineral compositions of selected MAD samples were analyzed by X-ray powder diffraction (XRD) and a computer software (SIROQUANT) based on Rietveld quantification method. For the XRD analysis, a Philips X'pert MPD diffractometer was used with a CuKa radiation in conditions of 40 kV and 20 mA.

Shear strength: A Torvane device was used to measure shear strength near the exposed sediment surface of split cores. This device is operated by inserting adapters 5 mm into the exposed sediment surface. The top of the spring-loaded Torvane is rotated thereby producing a torque that shears the sediment. A pointer records the maximum torque value, which is proportional to the shear strength. The Torvane comes in 3 diameters, 19, 25 and 48 mm, which measure a maximum shear stress up to 20, 100 and 250 kPa, respectively. Each size records on a continuous scale of 0–10 units, and measurements are multiplied by approximately 2, 10, and 25, respectively to obtain shear strength in units of kPa.

Thermal conductivity: A needle probe system was used, consisting of a heating wire and a thermocouple installed within a 1 mm diameter metal needle. Heat is generated by imposing a DC current through the heating wire, while the temperature evolution within the needle is monitored using the thermocouple: the higher the thermal conductivity of the medium, the higher the rate of heat dissipation and the lower the rate of temperature increase detected with the thermocouple. The electric current I is related to the voltage dropV_{ref} across a reference resistor R_{ref} placed in series with the heating wire.

$$I = \frac{V_{ref}}{R_{ref}}$$

(1)

Then, the input power Q is

$$Q = I^2 \cdot R_m = \left(\frac{V_{ref}}{R_{ref}}\right)^2 \cdot R_m$$

(2)

where R_m is the resistance of the heating wire. The early portion of the temperature time series is affected by the needle–sediment coupling while specimen boundaries perturb the long-time data. Therefore, the thermal conductivity is obtained from the linear, central portion of the temperature vs. log time plot. The thermal conductivity is computed as (derivation in Carslaw and Jaeger, 1959).

$$k = \frac{Q}{4\pi} \cdot \frac{\ln\left(\frac{t_2}{t_2}\right)}{T_2 - T_1} = \left(\frac{V_{ref}}{R_{ref}}\right)^2 \cdot \frac{R_m}{4\pi} \cdot \frac{\ln\left(\frac{t_2}{t_2}\right)}{T_2 - T_1}$$

(3)

This methodology is valid for homogeneous, isotropic materials (details can be found in Manohar et al., 2000 and ASTM, 2000).

RESULTS AND DISCUSSIONS

General Porosity, Density, and Grain Size Trend

Porosity, one of the basic physical properties, is affected by fabric, particle shape, mineralogy, pore water composition, and geological processes such as sedimentation rates, consolidation processes, and diagenesis (Winters et al., 2008). The porosity measured from cores ranges from 40% to 90% and the averaged value at each site ranged from 65% to 71% in the Ulleung Basin (Fig. 2 and Table 1). These values are relatively high in comparison with other marine gas hydrate research sites such as in the Gulf of Mexico, Cascadia Margin, and Nankai Trough (Kinoshita et al., 2009, Riedel et al., 2006a and Winters et al., 2008). The porosities measured from cores are sometimes higher than in-situ values due to disturbances induced by gas hydrate dissociations and degassing. To prevent this kind of error, MAD samples were taken from good quality cores, avoiding soupy layers produced from gas hydrate dissociations and severely cracked intervals by degassing. The porosity values measured directly from cores and those estimated from LWD log matches well each other. Consequently, this relatively high porosity trend is not induced from core disturbances but originated by inherent sediment characteristics. Other gas hydrate areas are mostly near conventional oil and gas field and the sediments sources are mostly fluvial in nature whereas the sediments in the Ulleung Basin were originated from mass transport deposits that were evolved from repeated slope failures (Bahk et al., 2013). For the origin of sediments, a large portion of bio-originated material, which has high internal porosity, are present in sediments and serve as major cause for high porosity values. The high OPAL-A content in mineral composition also supports this observation (Table 2). The sediments with high internal porosity produce high specific surfaces and behave very differently from those with similar grain size and no internal pores, showing a lower elastic wave velocity, a higher electrical surface conduction, a higher plasticity, and a lower permeability (Lee et al., 2008, Lee et al., 2010a, Lee et al., 2010b and Lee et al., 2010c). Consequently, caution should be taken in the interpretation of geophysical logging when such material is abundantly present. Generally, the porosity decreases rapidly with depth in shallow regions and decreases moderately afterward, suggesting that the shallow sediments at some sites have not reached normally consolidated condition. At shallow depths, due to the low vertical effective stress, bounded water in clays as controlled by surface charges and chemical effects can dominate mechanical consolidation, preventing normal consolidation (Francisca et al., 2005). A depth related porosity law, Athy's law, is a good tool for quantitatively assessing depth–porosity trends. The differences in the compaction parameter of Athy's law can be explained by both mineralogy differences and sedimentation rates.

The compaction parameter fitted to data in the Ulleung Basin ranges from 0.0003 (UBGH2-2_2) to 0.0017 (UBGH2-3) m^{-1} and averages to 0.0008 m^{-1}. The compaction parameters at chimney sites (UBGH2-2_1, 2-3, 2-7, and 2-11) are relatively higher, ranging from 0.0009 to 0.0017 m^{-1}. Those for other sites lie between 0.0005 and 0.0006 m^{-1}, except for Sites UBGH2-2_2 and 2-6, where the values are 0.0003 and 0.0008 m^{-1}, respectively. The bulk density and the grain density used for porosity estimation from all sites are available in the Supplementary data online.

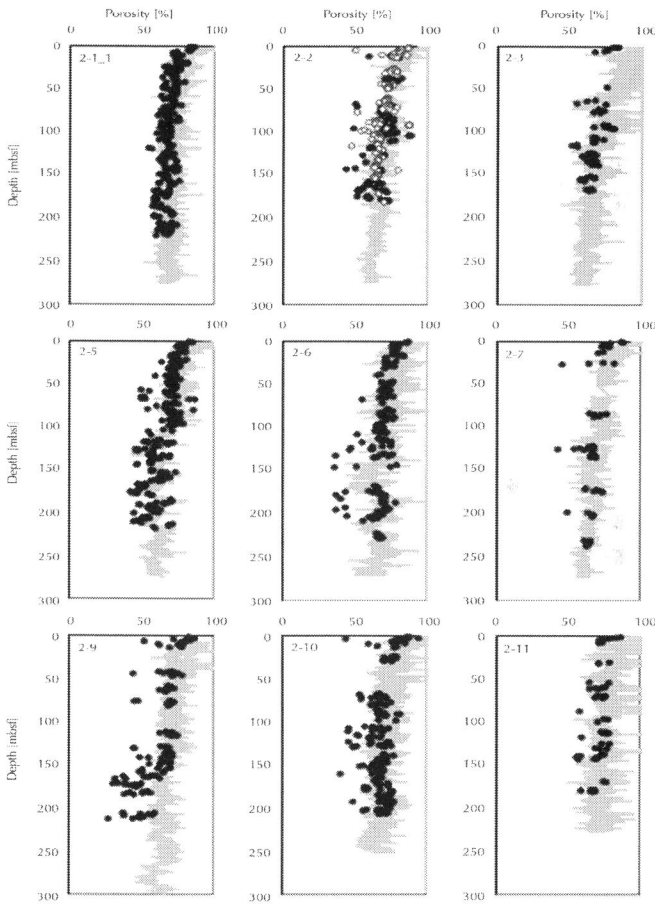

Figure 2: The porosity as a function of depth. Black dots and hollow dots are porosity values measured from cores and gray lines are porosity values estimated from LWD log. Sites UBGH2-2_2 and UBGH2-2_1 are plotted

together for comparison. Hollow dots indicate data from UBGH2-2_1 and black dots indicate those from UBGH2-2_2.

Table 1: Averaged porosity, grain density, and grain size at sites

Site	Porosity [%]	Grain density [g/cm3]	Grain size [μm]
2-1-1	70	2.61	9
2-2-1	70	2.57	16
2-2-2	69	2.65	17
2-3	68	2.62	14
2-5	65	2.66	20
2-6	68	2.64	17
2-7	68	2.66	14
2-9	69	2.61	19
2-10	69	2.58	16
2-11	71	2.61	15

Table 2: Averaged XRD mineral compositions for sites cored during UBGH2

Site	Quartz Albite K-feldspar [%]	Opal -A [%]	Calcite [%]	Muscovite [%]	Dolomite [%]	Chlorite Kaolinite Illite [%]	Pyrite [%]
2-1_1	31.3	30.2	4.6	11.7	0.1	17.8	2.6
2-2_2	35.8	28.1	6.4	10.6	0.2	16.0	2.4
2-5	34.1	27.8	4.2	11.3	0.2	18.6	2.4
2-6	37.8	26.9	3.5	10.8	0.1	16.2	2.8
2-10	32.4	34.2	5.2	9.6	0.0	14.8	2.5

The averaged grain density at each site ranges from 2.57 to 2.66 g/cm^3 and do not show any significant trend with depth. The values are relatively lower than other marine GH bearing sites due to the similar reason for high porosities (Kinoshita et al., 2009, Riedel et al., 2006a and Winters et al., 2008). The bio-originated, or diatomaceous, sediments with high internal porosity are abundantly present in the Ulleung Basin (Kwon et al., 2011 and Lee et al., 2011a). The opal-A content from XRD analyses indicates the content of diatomaceous sediments and the negative relation with opal-A content, and the grain density reported in Bahk et al. (2013) also supports this explanation.

The grain size distribution of sediments plays an important role in the gas hydrate occurrence by determining the occurrence types such as grain-displacing and pore-filling types and also the abundances from controlling the diffusion of methane gas. The grain size analyses of sediments from the Ulleung Basin shows that D_{50} (diameter at 50% of cumulative grain size curve), the mean grain size, mostly lies between 4 μm and 25 μm, which corresponds to those for clay and medium silt, with some exceptions at some intervals (Fig. 3). Site UBGH2-1_1 shows more or less constant D_{50}, whereas all other sites have scattered coarse grained intervals. Sites UBGH2-2_1, 2-2_2, 2-10, 2-9, and 2-11 are closely located and are also close to Site UBGH1-9 where hydrate-bearing sand body identified during the UBGH1 drilling expedition. Sites UBGH2-2_1, 2-2_2, and 2-10 show slight increases in D_{50} at the deeper depth with frequent coarse grained intervals at Sites UBGH2-2_2 and 2-10. At Sites UBGH2-2_2 and 2-10, hydrate-bearing thin sand layers also have been found in a significant number of intervals (Ryu et al., 2012). Sites UBGH2-9 and 2-11 do not show significant coarse grained interval although they are located near Sites UBGH2-2_2 and 2-10. Site UBGH2-6 is the most northern site and shows most thick coarse grained intervals that also contain significant amount of pore-filling gas hydrate, according to Lee et al. (2013).

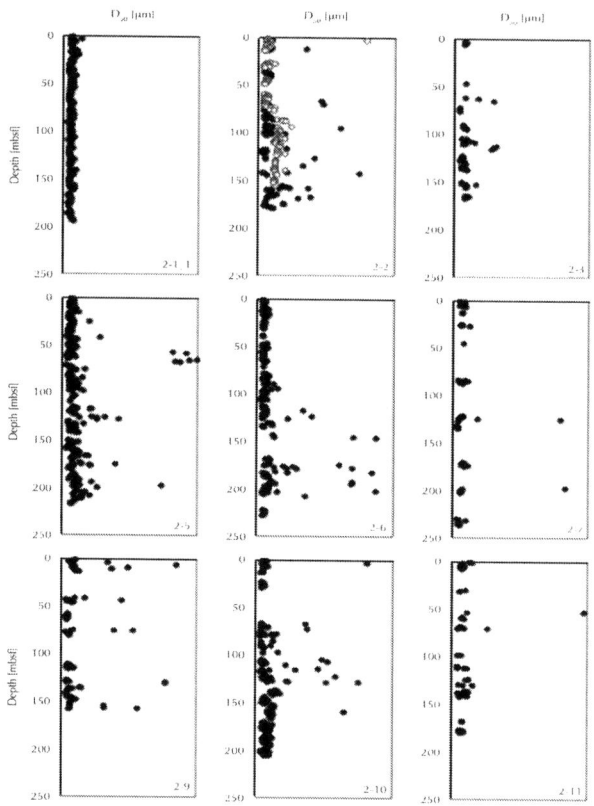

Figure 3: Mean grain size as a function of depth. Black dots and hollow dots are porosity values measured from cores and gray lines are porosity values estimated from LWD log. Sites UBGH2-2_2 and UBGH2-2_1 are plotted together for the comparison. Hollow dots indicate data from UBGH2-2_1 and black dots indicate those from UBGH2-2_2.

The fabric, or the spatial arrangement of sediments, is one of the factors that influence porosity and is affected by both the gravimetric force and surface electrical forces. The coarser particles are mostly affected by gravimetric forces, and the finer particles with higher specific surfaces are mostly affected by contact electrical forces. However, even very fine sediments are inevitably affected by gravimetric forces at deeper depths with higher vertical effective stress.

In greater depths, where the high vertical effective stress dominates the surface electrical forces, the sediments with the wider range of grain size have lower porosity, since the pore spaces of coarse sediments can be filled with finer ones resulting in lesser porosities. In the Ulleung Basin, D_{50}, mostly lies between 4 µm and 25 µm, as mentioned earlier, so more fraction of coarse grains will result in lower porosities. Overall, the porosity is inversely proportional to D_{50} when the porosity is smaller than a certain threshold, as predicted, but shows a negligible or slight positive trend to D_{50} when the porosity is larger than the threshold (Fig. 4). The threshold value slightly varies among sites but generally lies around 60% in the Ulleung Basin. This threshold could be explained with a couple of main causes. First, the relation between D_{50} and the OPAL-A content, an indicator of diatomaceous sediments, is responsible for the threshold. As mentioned earlier, the diatomaceous sediments have high internal porosities, so the high OPAL-A fraction results in high porosity. The OPAL-A fraction increases with the decrease in D_{50} until D_{50} reaches around 25 µm and shows negligible trend afterward, suggesting that the grain size of diatomaceous sediments is mostly less than 25 µm (Fig. 4). As a result, the negative relation between D_{50} and the porosity also become negligible when D_{50} is less than around 25 µm. Secondly, the threshold is formed because the fabric and surface electric forces interaction dominates rather than gravimetric sorting effects in fine sediments. The mean grain sizes (D_{50}) of sediments with the porosity larger than 60%, are mostly smaller than 25 µm, implying that the sediments with the porosity over 60% are mostly in surface electric force dominated sediments.

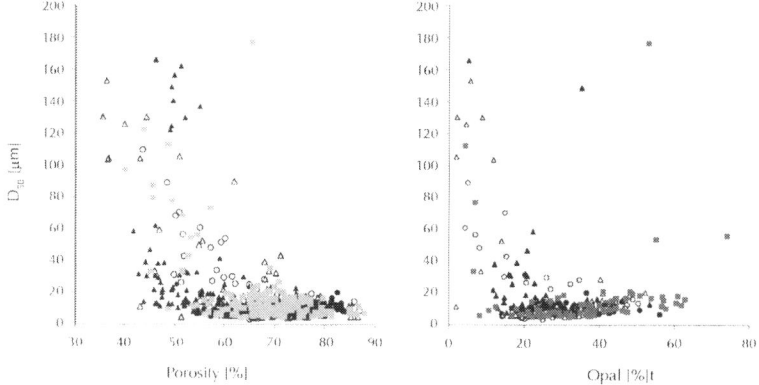

Figure 4: The trend of the mean grain size, D_{50}, as functions of porosity and opal fraction. Black circles are UBGH2-1_1, hollow circles are UBGH2-2_2,

black triangles are UBGH2-5, hollow triangles are UBGH2-6, and gray rect-angles are UBGH2-10.

Undrained Shear Strength

The undrained shear strength, S_u, is considered for situations when loading is applied so rapidly that excess pore water pressure cannot dissipate. Most fine grained sediments have low permeability and generally are involved with undrained conditions. In undrained conditions, the safety issues become critical immediately after an engineering loading, when the excess pore water pressure is the highest. The undrained shear strength is a key strength parameter for designing well completion in fine-grained sediments together with the unit weight of sediments while the friction angle and the unit weight are mainly considered in coarse grained sediments. The use of simple devices in the measurement of undrained shear strength, such as a hand held torvane and a pocket penetrometer gives crude assessments but has statistical advantages, offering numerous measurement points.

The S_u value of fine sediments ranges widely from almost zero for very soft sediments to a few MPa for very stiff sediments. In the Ulleung Basin, S_u values start from a few kPa at shallow depths and increase with depth up to a couple of hundred kPa. The undrained shear strength generally increases with depth since the bulk density, which highly affects S_u, increases with depth due to the compaction by the overburden pressure. The trends of S_u with depth, bulk density, and porosity in the Ulleung Basin are plotted inFigure 5. Since the trend shows higher scatter with greater depth, the trend lines were estimated conservatively, to avoid the overestimation of S_u. The immediate settlement estimation of fine grained sediments involves the undrained modulus E_u, and the ratio, E_u/S_u, ranges from 500 to 1500 in most cases (Bjerrum, 1972). According to the estimation by Bjerrum (1972), E_u ranges from a few hundred kPa near seafloor up to a few hundred MPa near 200 mbsf.

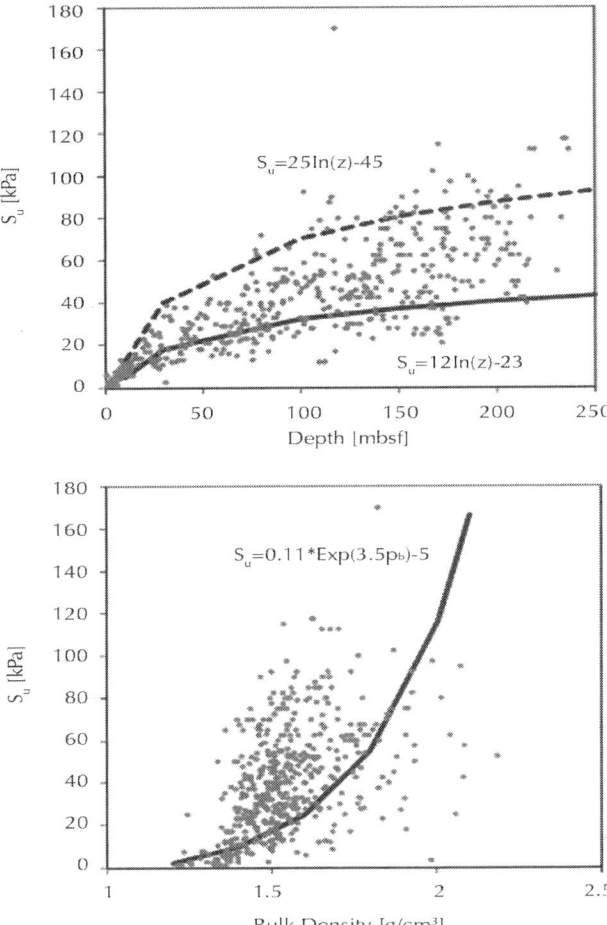

Figure 5: The trends of undrained shear strength functions of depth and bulk density in the Ulleung Basin.

The S_u value is sometimes normalized with the vertical effective stress and compared to other parameters (Holtz and Kovacs, 1981). The normalized undrained shear strength, $S_u/'_v$, is also called c/p ratio and is found to increase with the plasticity index by Skempton and Henkei (1953). However, it is also strongly affected by the stress history of the sediments (Bjerrum, 1972 and Ladd et al., 1977). For instance, the c/pratio of overly consolidated sediments is higher than that of normally consolidated sediments, and the sediment is usually considered overconsolidated when c/p ratio is greater than 0.5 (Hunt, 1984). The c/pratios are mostly less than 0.5 except some points at shallow depth, suggesting that the sediments have been normally consolidated under the present overburden pressures. The c/p ratio is relatively high near the seafloor and rapidly decreases with depth until ~10 mbsf (Fig. 6). The S_u range at shallow depth is 0.1–0.6 and that at deeper depth is 0.02–0.3. The phenomenon manifested by high c/p ratio near the seafloor is also called 'apparent overconsolidation' and has been reported also in other areas (Collet et al., 2008,Riedel et al., 2006a and Riedel et al., 2006b). Overconsolidation or apparent overconsolidation occurs due to 1) creep or secondary consolidation, 2) cementation, 3) cyclic freezing and thawing, 4) surface erosion, 5) slow sedimentation or rapid pore water drainage and 6) Thixotrophy (Johns, 2007, Poulos, 1988 and Lee et al., 2010c). Diatomaceous material is also reported as a cause of apparent overconsolidation (Johns, 2007). Sediments with high specific surfaces, like diatomaceous sediments, tend to have high cohesion (i.e. high internal strength), resulting in high c/p ratio at shallow depths (Skempton, 1970). In general, sediments are overconsolidated in the continental shelves and slopes due to slow sedimentation and surface erosion (Fukuoka and Nakase, 1973 and Poulos, 1988). In hydrate-bearing sediments, cyclic formation and dissociation of gas hydrate may induce overconsolidation. However, most of the causes listed above induce both low porosity and high stiffness, which is not the case in the Ulleung Basin, where the porosity is abnormally high as if the normal consolidation has not been completed. The most probable causes of the apparent overconsolidation in the Ulleung Basin are the surface erosion or thixotrophy. Thixotrophy occurs mostly in fine-grained sediments and in flocculated-prone conditions (Santamarina et al., 2001 and Van Olphen, 1951) Fine-grained sediments tend to flocculated in high-ionic condition such as marine environments (Mitchell, 1993). The ionic concentration is usually the highest at the seafloor and gradually dilutes, so more flocculation can occur near the seafloor. Unsaturated conditions such as gassy sediments also cause thixotrophy hardening by capillary effects (Diaz-Rodriguez and Santamarina, 2001).

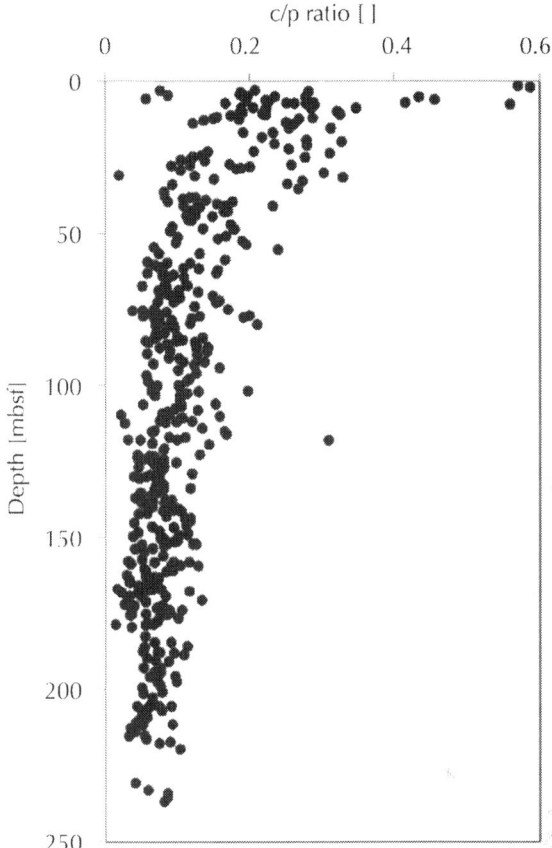

Figure 6: The ratio of undrained shear strength to the vertical effective stress, c/p ratio, as a function of depth.

Compression Index

The sediment compressibility is one of the key properties for assessing the borehole stability and the permeability evolution during the gas production from hydrate-bearing deposits. The sediment compressibility is characterized by the compression index C_c. The compression index C_c is defined as the slope of the compression line on a plot of void ratio vs. the log of vertical effective stress, and can be computed as follows;

$$C_c = -\frac{e_2 - e_1}{\frac{\log(\sigma'_v)_1}{(\sigma'_v)_2}}$$

(4)

where void ratio, e, is related to porosity n through e = n/(1 − n), and subscripts 1 and 2 indicate arbitrary two points at the compression line.

The compression index values for drilling sites can be roughly estimated using Equation (4) and the MAD data. The void ratio, e, can be calculated from the water content of cores as $G_s W = Se$, where G_s is the specific gravity of a grain, W is the gravimetric water content, S is the degree of saturation, and e is the void ratio. In marine sediments, the degree of saturation is 1, unless the area has a free gas phase. Vertical effective stress values have been calculated using bulk density from cores, assuming hydrostatic pore pressure. Hydrostatic pore pressure is assumed since no significant over or underpressured zones were detected during the drilling (Ryu et al., 2012).

In case of commercially available laboratory sediments, C_c is around 0.2 for kaolin clays and 0.02 for Ottawa sand (Lee et al., 2010a). The C_c values range from 0.4 to 1.1 in the Krishna–Godavari Basin in India, from 0.2 to 0.7 in the Cascadia margin in Canada and from 0.3 to 1.1 in the Gulf of Mexico, USA (Collet et al., 2008, Riedel et al., 2006a and Winters et al., 2008). The estimated compression index in the Ulleung Basin ranges from 0.5 to 1.5, which is slightly higher than those from other GH-bearing regions (Table 3). Sites UBGH2-3 and 2-11 are seismic chimney sites and contain a large amount of grain-displacing GH in pre-formed or induced fractures (Lee et al., 2013 and Ryu et al., 2012). The significantly low C_c values in these sites may have resulted not from actual sediment properties such as large D_{50} or low specific surfaces, but from porosity disturbances due to the extensive dissociation of grain-displacing gas hydrate that presents as a fracture-like structure. Some sites have two distinctively different C_c values within one site, where the C_c values are higher at shallow depths and lower at deeper depths. The depth that divides the C_c trend for Sites UBGH2-2_1, 2-7, and 2-10 are 5, 3, and 6 mbsf, respectively. The high C_c at shallow depth has been induced by the same reasons that caused abnormally high porosity near the seafloor stated in Section 4.1 of this report.

Table 3: The estimated compression index with Equation (4). The C_c values with * indicate where C_c has been altered by the extensive dissociation of grain-displacing GH in sediments. The sites with C_c in parenthesis indicate sites that have two distinctively different C_c in one site. The first value in parenthesis indicates C_c at shallow depth, and the second one indicates one at deeper depth. The depth that divides the e–log P trend for UBGH2-2_1, 2-7, and 2-10 are 5, 3, and 6 mbsf, respectively

Site	Cc
2-1-1	1.1 (1.8, 0.9)
2-2-1	1.3
2-2-2	1.1
2-3	0.5*
2-5	1.4
2-6	1.1
2-7	0.9 (5.9, 0.7)
2-9	1.5
2-10	1.0 (2.5, 0.8)
2-11	0.6*

Thermal Properties

Thermal settings and properties not only govern the occurrence of gas hydrate but also affect production strategies. For example, the depressurization should take account gas hydrate reformation, and the thermal stimulation may not be a good option when thermal conductivity is very low.

Measured temperatures and estimated thermal gradients are tabulated at Table 4. The estimated thermal gradients are between 90 and 110 °C/km, except for Site UBGH2-7 where temperature was higher by about 10 °C than at similar depths at other sites in the Ulleung Basin. The Yamamoto Basin and the Japan Basin, adjacent basins in Japan, show similar thermal gradients, thermal conductivities, and heat flows (Langseth and Tamaki, 1992). These relatively high thermal gradients in the Ulleung Basin imply that relatively high temperature conditions would be encountered at relatively shallow producing intervals, which could reduce the gas hydrate reformation during the production using the depressurization method. High thermal gradients in the Ulleung Basin are the product of both low thermal conductivity and high heat flow in the Ulleung Basin, which are discussed below.

Table 4: Temperature measurements acquired during the UBGH2 expedition

Site	Lat.	Long.	Water depth [mbss]	Depth [mbsf]	Temp [°C]	Thermal gradient [C/km]	Heat flow [mW /m2]
2-1-1	36.25	130.06	1534	0	0.2	100	57
				163.5	16.5		
				181.5	18.5		
2-2-1	36.7	130.89	2092	0	0.3	108	87
				156.0	17.6		
				190.0	20.5		
2-2-2	36.71	130.89	2093	0	0.2	107	83
				171.0	18.8		
				194.0	20.8		
2-3	35.69	130.34	898	0	0.3	95	55
				120.5	11.7		
2-5	36.52	130.56	1974	0	0.5	96	49
				147.5	15.5		
				210.0	19.6		
2-6	37.02	130.26	2153	0	0.5	112	65
				128.0	15.5		
				189.0	20.8		
2-7	36.92	130.36	2145	0	0.4	171	128
				168.0	29.8		
				204.0	34.7		
2-9	36.79	131.06	2106	0	0.2	115	65
				170.0	19.8		
2-10	36.93	130.9	2144	0	0.2	115	65
				113.5	11.7		
				168.0	19.8		
				189.5	21.9		

2-11	36.67	130.9	2082	0	1.2	112	73
				61.5	9.8		
				107.0	13.2		
				170.5	19.8		

The thermal conductivity of sediments is mostly governed by bulk density and mineral compositions, in addition to sediment fabrics. Thermal conductivity is generally considered to be proportional to the bulk density (or porosity) but is actually the result of some interplay among those factors above. The measured thermal conductivities average around 0.8 W/mK. Although measurements may have been biased toward low values due to core disturbances and preferential recovery of muds (Harris et al., 2011), the in-situ measurements of thermal conductivity in the Ulleung Basin from a previous work also ranges from 0.82 to 0.95, showing values below 1.0 W/mK (Kim et al., 2010). The high porosity and the abundance of clay mineral and OPAL-A in sediments are considered to be the main reason for this low thermal conductivity. The relatively low thermal conductivity in the Ulleung Basin suggests that thermal stimulation may not be a good option for gas hydrate production and could exacerbate the gas hydrate reformation problem by retarding heat flow from surrounding area to the cold gas hydrate dissociation front. For example, the thermal conductivity of calcite sediments with low porosity could be higher or lower than that of dolomite sediments with high porosity.

The heat flow, H, can be estimated with the temperature, T, and the thermal conductivity, k, as below.

$$T = T_0 + H \int_0^D \frac{dD}{k},$$

(5)

where T_0 is the bottom water temperature (Bullard, 1939 and Har4ris et al., 2011).

The estimated heat flows are higher near the center of the basin and lower near the continental shelf, except for that at Site UBGH2-7 where exceptionally high temperature was measured (Table 4). The overall trends are in agreement with previous estimates, except for Site

UBGH2-7 (e.g. Horozal et al., 2009 and Kim et al., 2010). The source of heat flow has been suggested as volcanic structures formed during the seafloor spreading stage or a simple magmatic intrusion during the post-rifting stage (Kim et al., 2010 and Lee et al., 1999).

IMPLICATION TO PRODUCTION

Grain Size Characteristics

The macro scale behavior of sediments during production operations, such as particle migration, fluid flow, and deformation behavior, is governed by particle characteristics, fabric formation environments, and the resulting fabrics. The geotechnical terms for characterizing the grain size distribution are useful tools for estimating geotechnical parameters in the absence of an extensive set of the geotechnical laboratory experiments. In this section, the grain size distributions have been utilized to characterize geomechanical behaviors induced by the production operations at Sites UBGH2-2_2 and 2-6, where pore-filling GH have been found in the intercalated turbidite sands.

The migration of particle during production can have serious consequences. During the second gas hydrate test production at the Mallik site of Mackenzie Delta, Canada, in 2007, the test was shut down since too much sand was produced and clogged the producing well (Yamamoto et al., 2008). Particle migration may results in subsurface erosion, inducing erosion tunnels and sinkholes (Terzaghi et al., 1996). The mean grain sizes of sediments from the Ulleung Basin are relatively finer than those of conventional oil and gas fields such as the Mallik site of Mackenzie Delta, where gas hydrate occurs at sand-dominated intervals (Yamamoto et al., 2008). At Sites UBGH2-2_2 and 2-6, most intervals are consisted of silts and clays and have some sandy layers. These sandy layers also contain fines, producing D_{50} of very fine sands. Furthermore, the thickness of those sandy layers are in cm-scale, so the intervals that can be isolated by packers should inevitably contain both sandy and silty layers, making the problem even more complicated. Various screen systems for preventing sand production during gas productions are commercially available, but the applications are mostly focused on lithified rock formations. Consequently, adopting particle migration concepts from soil-mechanics could be useful for assessing optimum screen system for production. Terzaghi et al. (1996) experimentally examined the filter requirements for preventing particle migrations and concluded that the filter size should be smaller than four times of D_{85} of fine sediments to prevent particle migration and larger than four times of D_{15} of coarse sediment to maintain proper permeability and thus seepage force, where D_{15} and D_{85} are the diameters at 15% and 85% of a cumulative grain size distribution curve. Consequently, a set of filters may be needed like an assemblage known as a graded filter to avoid both particle migrations and permeability reductions, when the grain sizes are widely distributed as in the Ulleung Basin (Terzaghi et al., 1996). Valdes and Santamarina (2008) suggested filter requirements by experimenting clogging and bridging concepts (Valdes and Santamarina, 2008). When the ratio of the screen size to the mean grain size is less than a certain threshold, stable bridge formations occur and prevent particle migration. The threshold ratio is different depending on particle shapes: ~3 for spherical particles and ~6 for platy particles.

The potential of particle migration can be checked with the grain size distribution (Lowe, 1988). The sediments can be considered as self-filtering, if the ratio of D_{15} of the coarser fraction to D_{85} of finer fraction is less than 4 or 5 with arbitrarily dividing the grain size curve into two parts. The possibility of self-filtering has been examined at Sites UBGH2-2_2 and 2-6, using the criteria above. At Site UBGH2-2_2, the ratios are less than 5 throughout all intervals, indicating that sediments at this site are self-filtering. At Site UBGH2-6, most intervals with fine grained sediments are self-filtering while the ratios at coarse-grained intervals suggest that the sediments are not self-filtering. These intervals correspond to those with D_{50} greater than ~35 μm (i.e. very coarse silt) and also with pore-filling gas hydrate. Consequently, the measures for controlling particle migration should be established when these coarse-grained intervals are to be produced. The ranges of suitable filter size according to Terzaghi et al. (1996) for Site UBGH2-6 are suggested in Figure 7. As fine-grained intervals are self-filtering, particle migration in these intervals should not be a significant problem even when a producing interval contains both fine-grained and coarse-grained layers. In this case, the high seepage forces in fine-grained layers due to low permeability would be the more problematic than the particle migration issues.

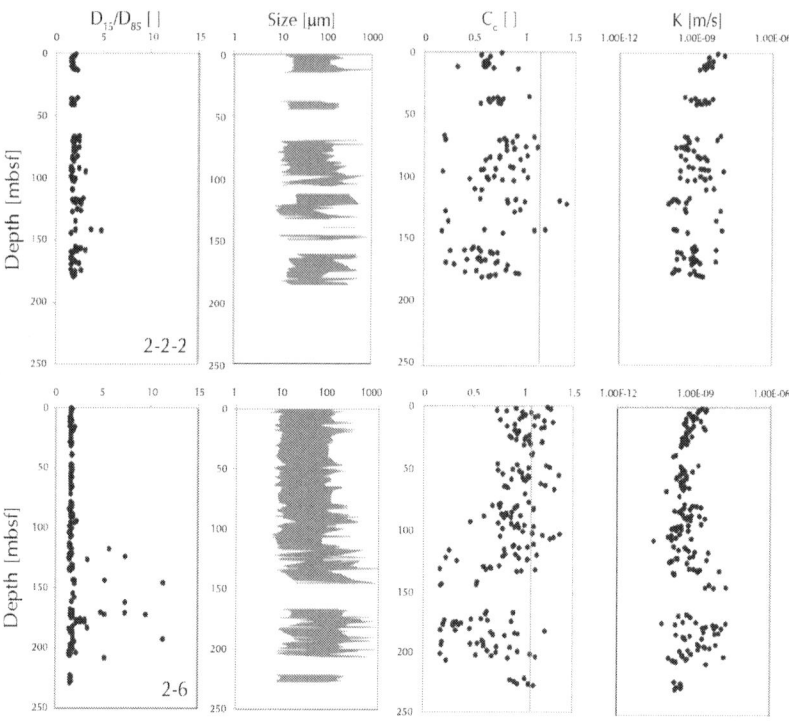

Figure 7: The characterization of particle migrations and the estimation of geomechanical properties based on particle characteristics. From the left, the ratio of D_{15} of the coarser fraction to D_{85} of finer fraction, the ranges of suitable filter size according to Terzaghi, the compression index estimated from the equations in Table 5, the hydraulic conductivity estimated from the equation in Table 5.

The cumulative grain size distributions are often used for an initial geotechnical characterization. The coefficient of uniformity, C_u, is the ratio of D_{60} to D_{10} and is very useful terms for empirically relating grain size distributions and geotechnical properties, where D_{10} and D_{80} are the diameters at 10% and 60% of a cumulative grain size distribution curve. When compared with terms "graded" and "sorted", high uniformity coefficients indicate well graded and poorly sorted sediments.

Specific surface, S_a, can be inferred from C_u for spherical particles (Table 5). Sediments with different C_u and same D_{50} have different specific surface distributions, and thus different geotechnical behaviors. Higher C_u values indicate a broader grain size distribution and more contributions of smaller particles to the total specific surfaces (Santamarina et al., 2001). Specific surface is a crucial parameter in geotechnical characterization, and numerous equations have been reported to relate the specific surface and other geotechnical properties (Table 5; Farrar and Coleman, 1967, Mitchell, 1993, Perloff and Baron, 1976,Terzaghi et al., 1996, Santamarina et al., 2001, Skempton, 1957 and Wood, 1990). According to these relations, high C_u indicates high specific surface, high plasticity index, low permeability, high compressibility, low pressure diffusion coefficient, low friction angle, and high undrained shear strength when normalized with the vertical effective stress. The compression index and the hydraulic conductivity estimated with the relations in Table 5 are plotted in Figure 7. The C_c values estimated from C_u range from 0.2 to 1.3 at Sites UBGH2-2_2 and 2-6. The C_c values estimated with porosity depth relation discussed in Section 4.3 are 1.2 and 1.1 for Sites UBGH2-2_2 and 2-6, respectively. The relation between C_c and the production-induced vertical deformation of gas hydrate deposits will be discussed later in Section 5.2 of this report. The hydraulic conductivity, K, governs production rates, the diffusion of depressurization front, and excess pore pressure dissipation rate. The usual range for producing interval for conventional gas production is 10^{-7}–10^{-6} m/s but the ranges are becoming wider and wider according to the advances in production techniques (Smith et al., 1992). Those for cap rocks are 10^{-10}–10^{-12} m/s. The estimated hydraulic conductivity at Site UBGH2-2_2 shows high and low variation throughout the interval since relatively thin sand layers frequently intercalated in mud throughout the interval (Bahk et al., 2013). Site UBGH2-6 has a relatively thick interval of high K at 110–200 mbsf, and there is ~20 m thick low K interval above the pore-filling GH-bearing interval (110–155 mbsf) which could serve as good seal during gas hydrate production.

Table 5: The empirical relationships used for estimating geomechanical properties related to the gas production from GH deposits. S_a is the specific surface; C_u is the coefficient of uniformity; ρ_w is the water density; D_{50} is the mean grain size; PI is the plasticity index; χ is the shape factor; γ_w is the unit weight of water; γ_g is the unit weight of the grain; η is fluid viscosity; e is the void ratio; g is the acceleration of gravity; C_c is the compression index; G_s is specific gravity; σ_y' is the effective stress; ø' is the peak friction angle; S_u is the undrained shear strength; K is the hydraulic conductivity (Farrar and Coleman, 1967, Mitchell, 1993, Perloff and Baron, 1976, Terzaghi et al., 1996, Santamarina et al., 2001, Skempton, 1957 and Wood, 1990)

Properties	Original expression
Specific surface	$S_a = \dfrac{3(C_u + 7)}{4\rho_w G_s D_{50}}$
Plasticity index	$PI = 0.5Sa - 8$
	$PI = 0.41Sa - 0.73$
Hydraulic conductivity	$K = \dfrac{1}{S_a^2}\dfrac{\chi \lambda_w g^2}{\eta \gamma_g^2}\dfrac{e^2}{(1+e)}$
Compressibility	$C_c = PI\dfrac{G_s}{200}$
	$C_c = \dfrac{PI}{74}$
	$Cc \approx 0.009(0.56 S2 + 9)$
Friction angle	$\sin\phi' = 0.35 - 0.1\ln(\dfrac{PI}{100})$
Undrained shear strength	$S_u = \sigma_v'(0.11 + 0.0037 PI)$
	$S_u = 0.22\sigma_v'$

Vertical Deformation Characteristics Induced by the Production

Although the dissociation of gas hydrate disturbs the sediment fabric and alters the C_c values, the settlement due to gas hydrate dissociation is highly dependent on C_c of sediments. The settlement during production is induced by the diminishing gas hydrate particle, pore pressure change, and seepage force, and the compression index can be utilized to assess the settlement due to these factors (Lee et al., 2010c). The estimation results, according to Lee et al. (2010c), in Figure 8 suggest that the vertical strain can be as high as 15% in pore-filling-GH bearing sandy layers in these sites (65–160 mbsf at Site UBGH2-2_2 and 110–155 mbsf at Site UBGH2-6). The subsidence induced by the pore pressure change from the depressurization is much higher than those induced by diminishing gas hydrate particles from gas hydrate dissociation. In fine-grained silty layers, massive vertical deformations have been predicted to occur due to the pore pressure change from the depressurization. The depressurization fronts in silty layers are unlikely to diffuse further enough to cause the regional deformation in a laterally large area. However, care should be taken in designing well completions since significant deformation could occur near well bore when a producing interval contains both sandy layers and silty layers.

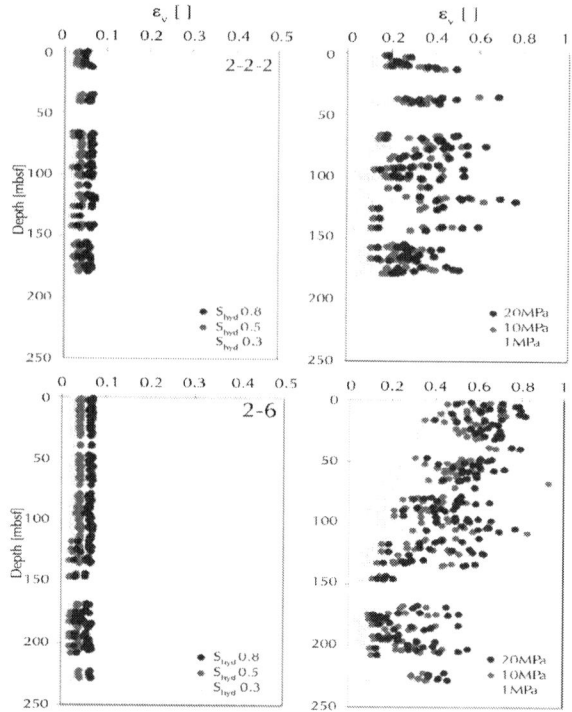

Figure 8: The estimation of vertical deformation induced from depressurization. Left: Deformation induced from diminishing hydrate particles, Right: Deformation induced from pore pressure changes.

Pore Pressure: Porosity Trend vs. Overpressure, In-Situ Pressures vs. Drilling Mud Pressure

Pore pressure is one of the important factors that control drilling operations. Overpressured situations are encountered during drilling operation more often than underpressured ones (Ingebritsen et al., 2006). Disequilibrium compaction, or undercompaction, is the most known cause of overpressure and occurs when a rapid compaction takes place on a permeable-sediment body isolated by low permeability sediments (Gordon and Flemings, 1998 and Swarbrick and Osborne, 1998). Similar compaction phenomenon occurs when a large scale tectonic compression takes place within a relatively short geologic time (Grollimund and Zoback, 2001 and Van Balen and Cloetingh, 1993). Dehydration reaction also causes an overpressure from the increase in the pore volume with excluded water from minerals (Alnes and Lilburn, 1998 and Bruce, 1984).

According to the geohazard assessment of the drilling sites based on seismic analysis, the possibility that formations are overpressured are negligible at shallow depths and generally low at deeper depths throughout the region (Fugro, 2010). Although thermal and geochemical analyses suggest the possibility of overpressure generation by dehydrate reaction, the presence of vertical fractures suggests that the generated overpressure has probably been dissipated through the fractures (Kim et al., 2013). Sometimes, the trend of porosity with depth is utilized to check the possibility of overpressured interval since the pore pressure change affects the vertical effective stress and thus the compaction trends (Burrus, 1998). The overpressure lowers the vertical effective stress and results in higher porosity than the normal porosity trend without abnormal pressure intervals. The porosity trends in the Ulleung Basin show no significant deviation from base line trends that indicates possible overpressured intervals. On the other hand, slight subnormal deviations are detected at some intervals at Sites UBGH2-6 and 2-5, suggesting possible underpressured intervals. However, when compared with grain density trend together, the abnormal grain density trends are mostly responsible subnormal porosity trends in these sites.

The pressures induced by drilling mud play a crucial role in maintaining stable well bore conditions and are generally applied at higher than the pore pressure and lower than the hydraulic fracture pressure, to prevent well bore collapse and hydraulic fracturing and thus to maintain well bore stability (Zoback, 2007). The hydraulic fracturing occurs when drilling mud pressure exceeds the minimum principal stress, so the hydraulic fracture pressure is determined by the minimum principal stress, σ_3. Eaton (1969) proposed the estimation of σ_3 with the vertical effective stress, σ'_v, the pore pressure, u, and the poison's ratio, v, as below:

$$\sigma_3 = \left(\frac{v}{1-v}\right)\sigma'_v + u$$

(6)

In deep sea drilling, the pore pressure induced by the water body above the seafloor is significant and determines the rough magnitude of the mud pressure required during drilling. At Sites UBGH2-2_2 and 2-6, water depths are 2092 m and 2154 m, so the hydrostatic pore pressure in the interval above 200 mbsf ranges from 21.0 to 22.8 MPa at Site UBGH2-2_2 and from 21.6 to 23.6 MPa at Site UBGH2-6 (Fig. 9). The total vertical stress ranges from 21.0 to 23.8 MPa at Site UBGH2-2_2 and from 21.6 to 24.0 MPa at Site UBGH2-6. The typical value of the Poisson's ratio is 0.2 for loose sands or silts, 0.4 for dense sands or silts, and 0.4–0.5 for saturated clays (Holtz and Kovacs, 1981). The sediments at Sites UBGH2-2_2 and 2-6 are either dense silts/sands or clays, so the Poisson ratio can be estimated around 0.4. The estimated σ_3 ranges from 21.0 to 23.4 MPa at Site UBGH2-2_2 and from 21.6 to 24.3 MPa at Site UBGH2-6. The Poisson's ratio tends to decrease with GH saturations (Lee et al., 2010a), so the σ_3 would become smaller in GH-bearing intervals, producing a narrower mud window. Note that the influence of GH on the Poisson's ratio from previous work is focused on pore-filing GH, and the influence of grain-displacing GH on Poisson's ratio would be different.

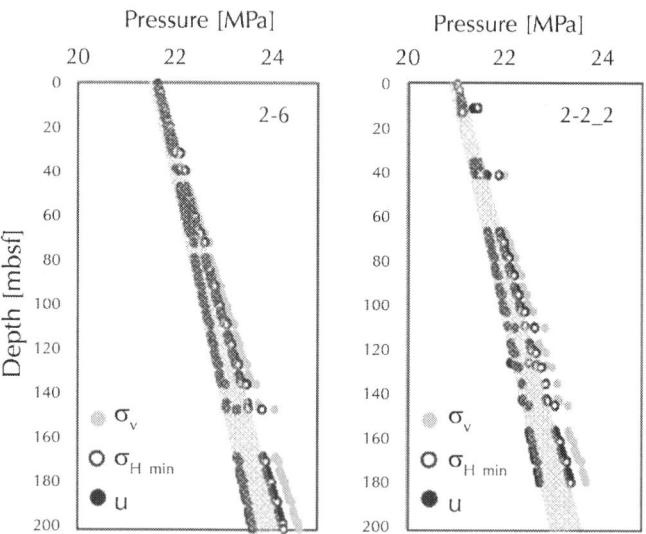

Figure 9: Total stress, σ_v, fracture pressure, σ_{Hmin}, and pore pressure, u, at sites UBGH2-2_2 and 2-6.

CONCLUSIONS

Physical properties, including index properties, undrained shear strength, and thermal conductivities were measured on conventional cores retrieved from the Ulleung Basin. The petrophysical and geomechanical analyses on these measurements reveal the following.

-The porosity ranges from 40% to 90%, and the grain density ranges from 1.8 to 2.9 g/cm^3. The main reason for the relatively high porosity and low grain density are the large portion of diatomaceous sediments.

-The mean grain size mostly lies between 4 μ m and 25 μ m (clay–medium silt), with sparsely scattered coarse grained intervals. The porosity is inversely proportional to D_{50} when the porosity is smaller than the threshold and shows a negligible or slight positive trend to D_{50} when the porosity is larger than the threshold. The threshold value slightly varies among sites but generally lies around 60%. The relation between D_{50} and OPAL-A contents is mainly responsible for this phenomenon.

-The undrained shear strength starts from a few kPa at shallow depths and increases with depth up to a couple of hundred kPa. The undrained shear strength generally increases with increasing depth and bulk density.

-The c/p ratios from all sites are mostly less than 0.5, except for some points at shallow depth, suggesting that the sediments have been normally consolidated under the present overburden pressures. The thixotrophy and surface erosion are most probable causes for the apparent overconsolidation at shallow depths in the Ulleung Basin.

-The compression index in the Ulleung Basin ranges from 0.5 to 1.5. The vertical deformation estimated from C_c suggests that the subsidence induced by the pore pressure change from the depressurization is much higher than those induced by the removal of gas hydrate particles by dissociations. Massive vertical deformations in fine-grained layers have been predicted to occur due to the pore pressure change from the depressurization.

-The thermal gradients mostly lie between 90 and 110 °C/km. The thermal conductivities average around 0.8 W/mK. The low porosity and the abundance of clay mineral and OPAL-A in sediments are the main reasons for this low thermal conductivity. The estimated heat flows are higher near the center of the basin and lower near the continental shelf and agree well with previous estimates, except for that at Site UBGH2-7 where exceptionally high temperature was measured.

-The macro scale behaviors of sediments during production operations are affected by particle characteristics and the resulting fabrics. According to the analyses on the grain size distributions, the particle migration in coarse grained layers during production necessitates proper measure for sand productions, while fine grained layers are mostly self-filtering. The hydraulic conductivity in pore-filling gas hydrate-bearing layers is relatively low in comparison with those from conventional gas reservoirs, and the conductivities at layers above GH layers are low enough to function as barriers.

-Neither significantly overpressured nor underpressured intervals are detected according to the porosity trends, which is also supported by the estimation from the structural analyses with seismic sections.

The drilling mud pressure operational window is mostly determined by the water depth in deep sea drilling, and caution should be taken in determining fracture pressure in hydrate-bearing sediments because gas hydrate saturations alter Poisson›s ratio.

ACKNOWLEDGMENTS

The author(s) wish to thank those that contributed to the success of the Second Gas Hydrate Drilling Expedition in the Ulleung Basin (UBGH2). Notably we wish to thank the co-chief scientists, the captains, crew and shipboard scientific party of the D/V Fugro Synergy. We also wish to acknowledge the support of the Gas Hydrate Research and Development Organization (GHDO) of the Ministry of Trade, Industry and Energy, Republic of Korea. Special appreciation is extended to Korea Institute of Geoscience and Mineral Resources, Korea National Oil Corporation, Korea Gas Corporation, Korea Ocean Research and Development Institute, Han Yang University, Korea Advanced Institute of Science and Technology, U.S. Geological Survey, Oregon State University, Geotek, Schlumberger, and Fugro Well Services who supported and participated in the 2010 UBGH2 drilling campaign. Samples and data for this study were acquired by onboard and post-cruise analyses of the Second Ulleung Basin Gas Hydrate Drilling Expedition (UBGH2), which was funded by Ministry of Knowledge Economy of Korea (MKE) under management of Gas Hydrate Research and Development Organization (GHDO) of Korea. We greatly thank the onboard scientific party and technical staff of UBGH2 for their support at sea.

REFERENCES

1. Alnes, J.R., Lilburn, R.A., 1998. Mechanisms for generating overpressure in sedimentary basins: a reevaluation. Discussion. Am. Assoc. Pet. Geol. Bull. 82, 2266e2269.

2. ASTM, 2000. Standard Test Method for Determination of Thermal Conductivity of Soil and Soft Rock by Thermal Needle Probe Procedure. Annual Book of ASTM Standards, pp. 8e9. D5334-00.

3. Bahk, J.J., Um, I.K., Holland, M., 2011. Core lithologies and their constraints on gashydrate occurrence in the East Sea, offshore Korea: results from the site UBGH1-9. Mar. Pet. Geol. 28, 1943e1952.

4. Bahk, J.J., Kim, G.Y., Chun, J.H., Kim, J.H., Lee, J.Y., Ryu, B.J., Lee, J.H., Son, B.K., Collett, T., Riedel, M., 2013. Characterization of gas hydrate reservoirs by integration of core and log data in the Ulleung Basin, East Sea. Mar. Pet. Geol. 47, 30e42.

5. Bjerrum, L., 1972. Embankments on Soft Ground. In: Proceedings of the ASCE Specialty Conference on Performance of Earth and Earth-supported Structures, vol. II. Purdue University, pp. 1e54.

6. Blum, P., 1997. Physical properties handbook: a guide to the shipboard measurement of physical properties of deep-sea cores. ODP Tech. Note 26. http:// dx.doi.org/10.2973/odp.tn.26.1997.

7. Bruce, C.H., 1984. Smectite dehydration: its relation to structural development and hydrocarbon accumulation in northern Gulf of Mexico basin. Am. Assoc. Pet. Geol. Bull. 68, 673e683.

8. Bullard, E.C., 1939. Heat flow in South Africa. Proc. R. Soc. Lond., Ser. A 173, 474e502. http://dx.doi.org/10.1098/rspa.1939.0159.

9. Burrus, J., 1998. Overpressure models for clastic rocks, their relation to hydrocarbon expulsion: a critical reevaluation e AAPG Memoir 70. In: Law, B.E., Ulmishek, G.F., Slavin, V.I. (Eds.), Abnormal Pressures in Hydrocarbon Environments. American Association of Petroleum Geologists, Tulsa, OK, pp. 35e63.

10. Carslaw, H.S., Jaeger, J.C., 1959. Conduction of Heat in Solids. Clarendon Press, Oxford, p. 510.

11. Chough, S.K., Barg, E., 1987. Tectonic history of the Ulleung Basin margin, East Sea (Sea of Japan). Geology 15, 45e48.

12. Chough, S.K., Lee, H.J., Yoon, S.H., 2000. Marine Geology of Korean Seas, second ed. Elsevier, Amsterdam, p. 313.

13. Collet, T., Riedel, M., Cochran, J., Boswell, R., Presley, J., Kumar, P., Sathe, A., Sethi, A., Lall, M., Sibal, V., The NGHP Expedition 01 Scientists, 2008. National Gas Hydrate Program Expedition 01 Initial Reports, DVD, p. 1828.

14. Diaz-Rodriguez, J.A., Santamarina, J.C., 2001. Mexico city soil behavior at different strains: observations and physical interpretation. J. Geotech. Geoenviron. Eng. 127 (9), 783e789.

15. Eaton, B.A., 1969. Fracture gradient prediction and its application in oilfield operations.

16. Journal of Petroleum Technology 10, 1353e1360.

17. Farrar, D.M., Coleman, J.D., 1967. The correlation of surface area with other properties of nineteen British clays. J. Soil Sci. 18, 118e124.

18. Francisca, F., Yun, T.S., Ruppel, C., Santamarina, J.C., 2005. Geophysical and geotechnical properties of near-seafloor sediments in the northern Gulf of Mexico gas hydrates province. Earth Planet Sci. Lett. 237, 924e939.

19. Fugro Geo Consulting Inc., 2010. Geohazard Site Assessment 2010 Gas Hydrate Exploration Program, the Ulleung Basin, Offshore Korea. Houston, Texas.

20. Fukuoka, M., Nakase, A., 1973. Problems of soil mechanics of the ocean floor. Proc. 8th Int. Conf. Soil Mech. Foundat. Engng 4 (2), 205e222.

21. Gordon, D.S., Flemings, P.B., 1998. Generation of overpressure and compactiondriven fluid flow in a Plio-Pleistocene growth-faulted basin, Eugene Island 330, offshore Louisiana. Basin Res. 10, 117e196.

22. Grollimund, B.R., Zoback, M.D., 2001. Impact of glacially-induced stress changes on hydrocarbon exploration offshore Norway. Am. Assoc. Pet. Geol. Bull. 87 (3), 493e506.

23. Harris, R.N., Schmidt-Schierhorn, F., Spinelli, G., 2011. Heat flow along the NanTroSEIZE transect: results from IODP expeditions 315 and 316 offshore the Kii Peninsula. Japan 12 (8), Q0AD16. http://dx.doi.org/10.1029/2011GC003593.

24. Holtz, R.D., Kovacs, W.D., 1981. An Introduction to Geotechnical Engineering. Prentice Hall, Englewood Cliffs, New Jersey, p. 733.

25. Horozal, S., Lee, G.H., Yi, B.Y., Yoo, D.G., Park, K.P., Lee, H.Y., Kim, W.S., Kim, H.J., Lee, K.S., 2009. Seismic indicators of gas hydrate and associated gas in the Ulleung Basin, East Sea (Japan Sea) and implications of heat flows derived from depths of the bottom-simulating reflector. Mar. Geol. 258, 126e138.

26. Hunt, R.E., 1984. Geotechnical Engineering Investigation Manual. McGraw-Hill Book Company, New York, p. 983.

27. Ingebritsen, S.E., Sanford, W., et al., 2006. Groundwater in Geologic Processes. Cambridge Press, Cambridge, United Kingdom.

28. Johns, M.W., 2007. Consolidation and Permeability Characteristics of Japan Trench and Nankai Trough Sediments from Deep Sea Drilling Project Leg 87, Sites 582, 583, and 584. http://dx.doi. org/10.2973/dsdp.proc.87.129.1986.

29. Jolivet, L., Shibuya, H., Fournier, M., 1995. Paleomagnetic rotation and the Japan Sea opening. In: Taylor, B., Natland, J. (Eds.), Active Margin and Marginal Basins of the Western Pacific, Geophys. Monogr., vol. 88. AGU, Washington, pp. 358e369.

30. Kim, Y.G., Lee, S.M., Matsubayashi, O., 2010. New heat flow measurements in the Ulleung Basin, East Sea (Sea of Japan): relationship to local BSR depth, and implications for regional heat flow distribution. Geo-Mar. Lett. 30, 595e603. http://dx.doi. org/10.1007/s00367-010-0207-x.

31. Kim, J.H., Park, M.H., Chun, J.H., Lee, J.Y., 2011. Molecular and isotopic signatures in sediments and gas hydrate of the central/ southwestern Ulleung Basin: high alkalinity escape fuelled by biogenically sourced methane. Geo-Mar. Lett. 31, 37e49. http:// dx.doi.org/10.1007/s00367-010-0214-y.

32. Kim, J.H., Torres, M.E., Hong, W.L., Choi, J.Y., Riedel, M., Bahk, J.J., Kim, S.H., 2013. Pore fluid chemistry from the second gas hydrate drilling expedition in the Ulleung Basin (UBGH2):

source, mechanisms and consequences of fluid freshening in the central part of the Ulleung Basin, East Sea. Mar. Pet. Geol. 47, 99e112.

33. Kinoshita, M., Tobin, H., Ashi, J., Kimura, G., Lallemant, S., Screaton, E.J., Curewitz, D., Masago, H.,H., Moe, K.T., The Expedition 314/315/316 Scientists, 2009. Proc. IODP, 314/315/316. Integrated Ocean Drilling Program Management International, Inc., Washington, DC http://dx.doi.org/10.2204/iodp.proc.314315316.2009.

34. Kwon, T.H., Lee, G.R., Cho, G.C., Lee, J.Y., 2011. Geotechnical properties of deep oceanic sediments recovered from the hydrate occurrence regions in the Ulleung Basin, East Sea, offshore Korea. Mar. Pet. Geol. 28, 1870e1883. http:// dx.doi.org/10.1016/j.marpetgeo.2011.02.003.

35. Ladd, C.C., Foote, R., Ishihara, K., Schlosser, Poulos, H.G., 1977. Stress-deformation and strength characteristics. State-of-the-Art Report. In: Proceedings of the Ninth International Conference on Soil Mechanics and Foundation Engineering, Tokyo, vol. 2, pp. 421e494.

36. Langseth, M.G., Tamaki, K., 1992. Geothermal measurements: thermal evolution of the Japan Sea basins and sediments. In: Tamaki, K., Suyehiro, K., Allan, J., McWilliams, M., et al. (Eds.), 1992. Proc. ODP, Sci. Results, vol. 127/128 (Pt. 2). Ocean Drilling Program, College Station, TX, pp. 1297e1309. http://dx.doi.org/10.2973/odp.proc.sr.127128-2.227.1992.

37. Lee, G.H., Kim, H.J., Suh, M.C., Hong, J.K., 1999. Crustal structure, volcanism and opening mode of the Ulleung Basin, East Sea (Sea of Japan). Tectonophysics 308, 503e525.

38. Lee, J.Y., Santamarina, J.C., Ruppel, C., 2008. Mechanical and electromagnetic properties of northern Gulf of Mexico sediments with and without THF hydrates. Mar. Pet. Geol. 25 (9), 884e895.

39. Lee, J.Y., Francisca, F.M., Santamarina, J.C., Ruppel, C., 2010a. Parametric study of the physical properties of hydrate-bearing sand, silt, and clay sediments: 2. Smallstrain mechanical properties. J. Geophys. Res. 115, B11105. http:// dx.doi.org/10.1029/2009JB006670.

40. Lee, J.Y., Santamarina, J.C., Ruppel, C., 2010b. Parametric study of the physical properties of hydrate-bearing sand, silt, and clay

sediments: 1. Electromagnetic properties. J. Geophys. Res. 115, B11104. http://dx.doi.org/10.1029/2009JB006669.

41. Lee, J.Y., Santamarina, J.C., Ruppel, C., 2010c. Volume change associated with formation and dissociation of hydrate in sediment. Geochem. Geophy. Geosyst. 11 (1). http://dx.doi. org/10.1029/2009GC002667.

42. Lee, C.H., Yun, T.S., Lee, J.S., Bahk, J.J., Santamarina, J.C., 2011a. Geotechnical characterization of marine sediments in the Ulleung Basin, East Sea. Eng. Geol. 117, 151e158. http://dx.doi. org/10.1016/j.enggeo.2010.10.014.

43. Lee, J.Y., Ryu, B.J., Yun, T.S., Lee, J.H., Cho, G.C., 2011b. Review on the gas hydrate development and production as a new energy resource. KSCE J. Civil Eng. 15 (4), 689e696. http://dx.doi. org/10.1007/s12205-011-0009-3.

44. Lee, J.Y., Jung, J.W., Lee, M.H., Bahk, J.J., Choi, J.Y., Kim, J.H., 2013. Pressure core based study on gas hydrate occurrences in the Ulleung Basin and implication to geomechanical controlling factors. Mar. Pet. Geol. 47, 85e98.

45. Liu, X., Flemings, P.B., 2007. Dynamic multiphase flow model of hydrate formation in marine sediments. J. Geophys. Res. vol. 112, B03101. http://dx.doi.org/ 10.1029/2005JB004227.

46. Lowe III, J., 1988. Seepage analysis. In: Jansen, R.B. (Ed.), Advanced Dam Engineering for Design, Construction, and Rehabilitation. Van Nostrand Reinhold, New York, p. 811.

47. Manohar, K., Yarbrough, D.W., Booth, J.R., 2000. Measurement of apparent thermal conductivity by the thermal probe method. J. Test. Eval. 28 (5), 345e351.

48. Masui, A., Miyazaki, K., Haneda, H., Ogata, Y., Aoki, K., 2008. Mechanical characteristics of natural and artificial gas hydrate bearing sediments. In: Paper 5697 Presented at the 6th International Conference on Gas Hydrates, Chevron, Vancouver, B.C., Canada, 6e10 July.

49. Mitchell, J.K., 1993. Fundamentals of Soil Behavior, second ed. John Wiley & Sons, New York, p. 437.

50. Moridis, G.J., Kowalsky, M.B., Pruess, K., 2007. Depressurization-induced gas production from Class 1 hydrate deposits. SPE Reserv. Eval. Eng. 10, 458e481. http://dx.doi.org/10.2118/97266-PA.

51. Nimblett, J., Ruppel, C., 2003. Permeability evolution during the formation of gas hydrates in marine sediments. J. Geophys. Res. 108 (B9), 2420. http://dx.doi.org/ 10.1029/2001JB001650.

52. Perloff, W.H., Baron, W., 1976. Soil Mechanics Principles and Applications. The Ronald Press Company, New York, p. 745.

53. Poulos, H.G., 1988. Marine Geotechnics. Routledge, p. 473. Riedel, M., Collett, T.S., Malone, M.J., The Expedition 311 Scientists, 2006a. Proceedings of the Integrated Ocean Drilling Program (DVD), vol. 311. Integrated Ocean Drilling Program Management International, Inc., Washington, DC http:// dx.doi. org/10.2204/iodp.proc.311.103.2006.

54. Riedel,M., Long, P., Liu, C.S., Schultheiss, P., Collett, T., ODP Leg 204 Shipboard Scientific Party, 2006b. Physical properties of near-surface sediments at southern Hydrate Ridge: results from ODP Leg 204. In: Tréhu, A.M., Bohrmann, G., Torres, M.E., Colwell, F.S. (Eds.), Proc. ODP, Sci. Results, 204 [Online]. Available from World Wide Web: http://www-odp.tamu.edu/publications/204_SR/104/104.htm.

55. Ruppel, C., 2000. Thermal state of the gas hydrate reservoir. In: Max, M.D. (Ed.), Natural Gas Hydrate: in Oceanic and Permafrost Environments. Kluwer Acad., Dordrecht, Netherlands, pp. 29e42.

56. Rutqvist, J., Moridis, G.J., 2007. Numerical studies of geomechanical stability of hydrate-bearing sediments. In: The Offshore Technology Conference, Am. Assoc. of Pet. Geol., Houston, Tex., 30 April to 3 May.

57. Ryu, B.-J., Kim, G.Y., Chun, J.H., Bahk, J.J., Lee, J.Y., Kim, J.-H., Yoo, D.G., Collett, T.S.,

58. Riedel, M., Torres, M.E., Lee, S.-R., The UBGH2 Scientists, 2012. The Second Ulleung Basin Gas Hydrate Drilling Expedition 2 (UBGH2). Expedition Report. KIGAM, Daejeon, p. 667.

59. Santamarina, J.C., Klien, K.A., Fam, M.A., 2001. Soils and Waves. Wiley, New York, p. 488.

60. Skempton, A.W., 1957. The planning and design of the new Hong Kong airport. Discussion. In: Proceedings of the Institute of Civil Engineers, London, vol. 7, pp. 305e307.

61. Skempton, A.W., 1970. The consolidation of clays by gravitational compaction. J. Geol. Soc. Lond. 125, 373e411.

62. Skempton, A.W., Henkei, D.J., 1953. The post-glacial clays of the thames estuary at Tillbury and Shellhaven. In: Proceedings of the Third International Conference on Soil Mechanics and Foundation Engineering, Zurich, vol. I, pp. 302e308.

63. Smith, C.R., Tracy, G.W., Farrar, R.L., 1992. Applied Reservoir Engineering, vol. 1. OGCI and Petroskills Publications, Tulsa, Oklahoma.

64. Swarbrick, R.E., Osborne, M.J., 1998. Mechanisms that generate abnormal pressures: an overview. In: Law, B.E., Ulmishek, G.F., Slavin, V.I., Tulsam, O.K. (Eds.), Abnormal Pressures in Hydrocarbon Environments, AAPG Memoir 70. American Association of Petroleum Geologists, pp. 13e34.

65. Terzaghi, K., Peck, R.B., Mesri, G., 1996. Soil Mechanics in Engineering Practice, third ed. John Wiley & Sons, New York, p. 549.

66. Valdes, J.R., Santamarina, J.C., 2008. Clogging: bridge formation and vibration-based destabilization. Can. Geotech. J. 45 (2), 177e184.

67. Van Balen, R.T., Cloetingh, S.A., 1993. Stress-induced fluid flow in rifted basins. Diagenesis and basin development. A.D. Horbury and A.G. Robinson. Tulsa. Am. Assoc. Pet. Geol. 36, 87e98.

68. Van Olphen, H., 1951. Rheological phenomena of clay sols in connection with the charge distribution on the micelles. Discuss. Faraday Soc. 11, 82e84.

69. Waite, W.F., deMartin, B.J., Kirby, S.H., Pinkston, J., Ruppel, C.D., 2002. Thermal conductivity measurements in porous mixtures of methane hydrate and quartz sand. Geophys. Res. Lett. 29 (24), 2229. http://dx.doi.org/10.1029/2002GL015988.

70. Waite, W.F., Stern, L.A., Kirby, S.H., Winters, W.J., Mason, D.H., 2007. Simultaneous determination of thermal conductivity, thermal diffusivity and specific heat in sI methane hydrate. Geophys. J. Int. 169, 767e774. http://dx.doi.org/10.1111/j.1365-246X.2007.03382.x.

71. Winters, W.J., Dungan, B., Collet, T.S., 2008. Physical properties of sediments from Keathley Canyon and Atwater Valley. JIP Gulf Mexico gas hydrate drilling program. Mar. Pet. Geol. 25, 896e905.

72. Winters, W., Walker, M., Hunter, R., Collett, T., Boswell, R., Rose, K., Waite, W., Torres, M., Patilg, S., Dandekar, S., 2011. Physical properties of sediment from the Mount Elbert Gas Hydrate Stratigraphic Test Well, Alaska North Slope. Mar. Pet. Geol. 28, 361e380.

73. Wood, D.M., 1990. Soil Behavior and Critical State Soil Mechanics. Cambridge University Pres, Cambridge, p. 462.

74. Yamamoto, K., Numasawa, M., Yasuda, M., Fujii, T., Fujii, K., Dallimore, S.R., Wright, J.F., Nixon, M., Imasato, Y., Cho, B., Ikegami, T., Sugiyama, H., Mizuta, T., Kurihara, M., Masuda, Y., 2008. Objectives and operation overview of the 2007 Jogmec/Nrcan/Auroroa Mallik 2L-38 gas hydrate production test. In: Proceedings of the 6th International Conference on Gas Hydrates, Vancouver, British Columbia, Canada, July 6e10. 2008.

75. Zoback, M., 2007. Reservoir Geomechanics. Cambridge University Press, New York, p. 461.

Assessing Worker Exposure to Inhaled Volatile Organic Compounds from Marcellus Shale Flowback Pits

Ry Bloomdahl[a], Noura Abualfaraj[b], Mira Olson[b], and Patrick L. Gurian[b]

[a]Drexel University School of Public Health, Environmental and Occupational Health Department, 3215 Market St, Philadelphia, PA, 19104, USA
[b]Drexel University College of Engineering, CAEE Department, 3141 Chestnut Street, Phsiladelphia, PA, 19104, USA

ABSTRACT

Natural gas drilling sites employing hydraulic fracturing present a potential source of inhalation exposure to volatile organic compounds (VOCs) via the use of flowback pits. These open-air pits are used as a means of storing flowback water, a waste product of hydraulic fracturing, and represent an understudied source of VOC exposure for workers. The objective of this study was to assess this worker exposure and the resulting health risks for 12 VOCs present in flowback water stored in such an open reservoir on a drilling site. Flowback pit VOC mean, 2.5 percentile, and 97.5 percentile concentrations were used to model aqueous phase concentrations, and two models of volatilization were applied to estimate flux to the gas phase. A mass-balance approach was used to estimate gas phase concentrations that were, in turn, used to estimate worker exposure. A literature review was performed to determine VOC health effects, exposure limits, and worker protection methods. Neither model demonstrated an increased risk of adverse effects due to subchronic exposure at the 2.5 percentile and mean concentration values for the 12 VOCs as indicated by hazard quotients, hazard indices, or excess lifetime cancer risks; however, 97.5 percentile hazard indices approached 1 in one model and did demonstrate unacceptable risks in the evaluation of limitations. Either model may apply to worker health assessment depending upon industry practice; however, differing weather conditions, industry practice, and the small number of VOCs evaluated necessitate further research regarding worker risks and health effects.

INTRODUCTION

The United States has growing national concerns regarding energy availability and energy independence, and the expansion into shale gas drilling using hydraulic fracturing has gained more attention in recent years, both for its benefits and for its risks. Limited research has been done on sources of inhaled hazards for on-site workers from flowback storage pits. This project aimed to address some of these concerns by researching this disposal technique in which flowback fluid is stored in open-air pits. From previous work flowback was shown to contain, among other chemicals, volatile organic compounds, known to be hazardous to human health (Abualfaraj et al., 2014 and ATSDR, 2012), and this project attempted to assess potential inhalation exposure levels and risks associated with 12 of these volatile organic compounds (VOCs).

Exposure Assessments

The exposure assessment involved 3 sets of calculations (see Appendix A for a more detailed description). For all formulas, a pressure of 1 atm and a temperature of 25 °C were assumed.

The first set of calculations involved the determination of flux values, air emissions, and air concentrations of the 12 VOCs. Flux is reported in g/m^{2-s} and is used to arrive at an air emissions value (g/s). The air emissions value is then used to determine the air concentration, presented here as $\mu g/m^3$.

In the second set of calculations, the air concentration values were converted to exposure concentration values (following the method of EPA, 2009).

In the third set of calculations, the exposure concentration values were used to calculate subchronic inhalation exposures and excess cancer risk values (again following the method of EPA, 2009). Subchronic inhalation risk is presented using individual chemical hazard quotients, which can be summed into an overall hazard index (EPA, 2011c). A hazard quotient greater than 1 is considered a possible reflection of unacceptable risk, as is a hazard index greater than 1 (EPA, 2009 and EPA, 2011c).

A general range for the upper limit for allowable excess cancer risk is between 10^{-4} and 10^{-6} (Title 40: Part 300, 1994). Excess cancer risks for each chemical are summed to determine a total excess cancer risks (EPA, 2009).

Sources of Data and Assumptions

Flowback concentrations for this analysis were based on a study by Abualfaraj et al. (2014) which assembled a database of approximately 35,000 observations.

Abualfaraj et al. (2014) defined 6 of the 12 VOCs discussed (1,1,2-trichloroethane, 1,2-dichloroethane, 1,2-dichloropropane, benzene, carbon tetrachloride, and vinyl chloride) as high priority because their mean concentrations exceeded their MCLs by approximately 10 times. The remaining 6 VOCs chosen for study also had relatively high detected concentrations compared to their MCLs.

Calculating Flux Values and Air Emissions Values

Flux values were calculated using two separate models: the Stagnant Two-Film Model described inSchwarzenbach et al. (1993) and a model outlined in AP 42, Fifth Edition, Chapter 4, Section 3, published by the EPA regarding evaporation loss sources (1998). The latter model accounted for the presence of a thin oil-film layer, while the former did not. Only the Stagnant Two-Film Model formulas and results are presented here. To view the formulas and results for the AP42 model, please refer to Bloomdahl (2014).

All calculations were performed in Microsoft Office Excel 2007. The general formulas for the Stagnant Two-Film Model are described below (see Appendix A also). See Appendix B for greater description of variables and assumptions in all formulas described.

Stagnant Two-film Model

$$\text{Flux} = \left(\frac{1}{(Z_w/D_w) + (Z_a/D_a K'_H)} \right) \left(C_w - \frac{C_a}{K'_H} \right)$$

This calculation was performed for each individual VOC. Z_w is the thickness in centimeters (cm) of the stagnant water layer at the interface with the air. D_w is the diffusivity of the VOC in water in cm^2/s. Z_a refers to the thickness of the stagnant air layer in centimeters (cm) at the interface between the water and the air. D_a is the diffusivity of the VOC in air in cm^2/s. K'_H is the unitless form of Henry's constant. C_w is the concentration in $\mu g/cm^3$ of the VOC in the flowback water, and C_a is the initial concentration in $\mu g/cm^3$ in air of the VOC. Flux is given in $\mu g/cm^{2-s}$. To convert the Stagnant Two-Film Model's flux values to air emissions values, each of the flux values was multiplied by the assumed area of the flowback pit (125 ft × 125 ft = 15,625 ft² = 1451.61 m²) to quantify the chemical air emissions of the pit.

Air Emissions = Flux × Area of Flowback Pit in M²

All air emissions values are reported in grams per second (g/s).

Calculating Air Concentrations

Air emission values were converted to air concentration values by dividing the air emissions values by the air flow at the pit:

$$\text{Air concentration} = (\text{Air emissions value for chemical})/(\text{air flow})$$

Air concentrations are presented in $\mu g/m^3$. Air flow is calculated by multiplying air velocity by the product of worker height and flowback pit width. The average worker height was assumed to be 1.67 m to maximize exposure by allowing the chemical to reach breathing zone height but no higher.

Estimating Exposure Concentrations and Hazard Quotients

To evaluate inhalation exposure and risk, the EPA's Risk Assessment Guidance for Superfund (RAGS) Volume I: Part F was used (EPA, 2009). The timeframe for exposure was assumed to be 8 h/day, 5 days/week, 9 months (39 weeks) per year for one year; Pennsylvania codes pertaining to flowback pits state that the pits should be closed within 9 months after completion of drilling (unless otherwise permitted) (Chapter 78, 2001). Given this timeframe, a subchronic exposure was assumed, and Equations (8) and (12) were used to calculate Hazard Quotients, respectively. Equations (6) and (11) were used to calculate Excess Cancer Risks.

$$
\begin{aligned}
\text{Exposure Concentration} &= \text{Equation 6 or Equation 8} \\
&= (\text{Concentration in Air} \\
&\quad \times \text{Exposure time} \\
&\quad \times \text{Exposure frequency} \\
&\quad \times \text{Exposure duration})/ \\
&\quad \text{Averaging time}
\end{aligned}
$$

Hazard Quotient $=$ Equation 12

$\quad\quad = $ Exposure Concentration/(Toxicity Value

$\quad\quad\quad \times$ 1000 µg/mg)

Excess Lifetime Cancer Risk $=$ Equation 11

$\quad\quad\quad = $ Inhalation Unit Risk

$\quad\quad\quad\quad \times$ Exposure Concentration

RESULTS

The results for the flux values and air emissions are described in Table 2. As previously mentioned only the results of the Stagnant Two-Film Model are presented here; the EPA AP42 Model demonstrated much lower values in all calculations (a factor of approximately 10,000–100,000).

Table 2: Flux and air emission values for 12 VOCs

Volatile organic compound	Flowback concentration (µg/cm³)	Flux (g/m²-s)	Air emission (g/s)
Xylenes (total)	0.39 (0.0026–5.9)	1.2×10^{-6}($7.7 \times 10^{-9} - 1.7 \times 10^{-5}$)	0.0017 ($1.1 \times 10^{-5} - 0.025$)
Vinyl Chloride	0.074 (0.005–0.5)	3.5×10^{-7}($2.3 \times 10^{-8} - 2.3 \times 10^{-6}$)	0.000503 ($3.4 \times 10^{-5} - 0.0034$)
1,1,1-Trichloroethane	0.047 (0.00025–0.5)	1.6×10^{-7}($8.4 \times 10^{-10} - 1.7 \times 10^{-6}$)	0.00023 ($1.2 \times 10^{-6} - 0.0024$)
Toluene	0.24 (0.00028–2.2)	8×10^{-7} ($9.3 \times 10^{-10} - 7.3 \times 10^{-6}$)	0.0012 ($1.3 \times 10^{-6} - 0.011$)
Carbon Tetrachloride	0.048 (0.00025–0.5)	1.6×10^{-7}($8.5 \times 10^{-10} - 1.7 \times 10^{-6}$)	0.00024 ($1.2 \times 10^{-6} - 0.0025$)
Benzene	0.124 (0.00051–1.4)	4.6×10^{-7}($1.9 \times 10^{-9} - 5.03 \times 10^{-6}$)	0.00067 ($2.7 \times 10^{-6} - 0.0073$)
Ethylbenzene	0.049 (0.00025–0.5)	1.5×10^{-7}($7.4 \times 10^{-10} - 1.5 \times 10^{-6}$)	0.00021 ($1.1 \times 10^{-6} - 0.0022$)
1,2-Dichloropropane	0.047 (0.00025–0.5)	1.5×10^{-7}($8.2 \times 10^{-10} - 1.6 \times 10^{-6}$)	0.00022 ($1.2 \times 10^{-6} - 0.0024$)
Styrene	0.049 (0.00031–0.5)	1.5×10^{-7}($9.3 \times 10^{-10} - 1.5 \times 10^{-6}$)	0.00022 ($1.3 \times 10^{-6} - 0.0022$)

1,2-Dichloroethane	0.051 (0.0003–0.5)	1.8×10^{-7} (1.1 × $10^{-9} - 1.8 \times 10^{-6}$)	0.00027 (1.6 × $10^{-6} - 0.0026$)
1,2,4-Trichlorobenzene	0.039 (0.00025–0.5)	1.04×10^{-7} (6.7 × $10^{-10} - 1.3 \times 10^{-6}$)	0.00015 (9.7 × $10^{-7} - 0.0019$)
1,1,2-Trichloroethane	0.047 (0.00025–0.5)	1.5×10^{-7} (7.9 × $10^{-10} - 1.6 \times 10^{-6}$)	0.00022 (1.1 × $10^{-6} - 0.0023$)

Values are presented as: Mean (2.5th percentile – 97.5th percentile).

Xylenes had the highest mean concentration, flux, and air emissions values. 1,2,4-trichlorobenzene had the lowest mean concentration, flux, and air emissions values.

Air Concentrations

Calculated air concentrations are presented in Table 3.

Table 3: Air concentrations for 12 VOCs

Volatile organic compound	Air concentrations ($\mu g/m^3$)
Xylenes (total)	5.9 (0.039–89)
Vinyl Chloride	1.8 (0.12–12)
1,1,1-Trichloroethane	0.81 (0.0043–8.6)
Toluene	4.04 (0.0047–37)
Carbon Tetrachloride	0.83 (0.0043–8.6)
Benzene	2.3 (0.0095–26)
Ethylbenzene	0.74 (0.0038–7.6)
1,2-Dichloropropane	0.79 (0.0042–8.4)
Styrene	0.75 (0.0047–7.7)
1,2-Dichloroethane	0.94 (0.0055–9.2)
1,2,4-Trichlorobenzene	0.53 (0.0034–6.8)
1,1,2-Trichloroethane	0.76 (0.00402–8.04)

Values are presented as: Mean (2.5th percentile – 97.5th percentile).

Xylenes and 1,2,4-trichlorobenzene had the highest and lowest air concentration values, respectively.

Exposure Concentrations, Hazard Quotients, Cancer Risks

Exposure concentrations, hazard quotients, hazard indices, and excess lifetime cancer risks are described below and in Table 4 and Table 5.

Table 4: Stagnant two-film model exposure concentrations and hazard quotients

Volatile organic compound	Exposure concentration (µg/m³)	Hazard quotient (unitless)
Xylenes (total)	1.4 (0.0093–21)	0.014 (9.3×10^{-5}–0.21)
Vinyl Chloride	0.42 (0.028–2.8)	0.0042 (0.00028–0.028)
1,1,1-Trichloroethane	0.19 (0.00102–2.04)	3.8×10^{-5} (2.04×10^{-7}–0.00041)
Toluene	0.96 (0.0011–8.9)	0.00019 (2.2×10^{-7}–0.0018)
Carbon Tetrachloride	0.197 (0.00103–2.05)	0.00197 (1.03×10^{-5}–0.0205)
Benzene	0.56 (0.0023–6.09)	0.0186 (7.6×10^{-5}–0.203)
Ethylbenzene	0.18 (0.0009–1.8)	0.00018 (9×10^{-7}–0.0018)
1,2-Dichloropropane	0.19 (0.00099–2.0)	0.047 (0.00025–0.5)
Styrene	0.18 (0.0011–1.8)	0.00018 (1.1×10^{-6}–0.0018)
1,2-Dichloroethane	0.22 (0.0013–2.2)	0.0032 (1.9×10^{-5}–0.031)
1,2,4-Trichlorobenzene	0.13 (0.00081–1.62)	0.0063 (4.05×10^{-5}–0.081)
1,1,2-Trichloroethane	0.18 (0.00096–1.9)	NA[a]

Values are presented as: Mean (2.5th percentile – 97.5th percentile).

NA – not able to be calculated. Only subchronic and chronic p-RfC screening values were available for 1,1,2-trichloroethane in the PPRTV supporting documents. These screening values were not used as the confidence in these values was considered low (EPA, 2014b).

Table 5: Exposure concentrations and excess lifetime cancer risks for 12 VOCs

Volatile organic compound	Exposure concentration (µg/m³)	Excess lifetime cancer risk (unitless)
Xylenes (total)	0.0101 (6.7×10^{-5}–0.15)	Not available[a]
Vinyl Chloride	0.00302 (0.000204–0.0204)	1.3×10^{-8} (9×10^{-10}–9×10^{-8})
1,1,1-Trichloroethane	0.0014 (7.3×10^{-6}–0.015)	Not available[a]
Toluene	0.0069 (8.1×10^{-6}–0.064)	Not available[a]
Carbon Tetrachloride	0.0014 (7.4×10^{-6}–0.015)	8.5×10^{-9} (4.4×10^{-11}–8.9×10^{-8})

Benzene	0.004004 (1.6 × 10⁻⁵–0.044)	2.002x10⁻⁸ (8.2 × 10⁻¹¹–2.2 × 10⁻⁷)
Ethylbenzene	0.0013 (6.5 × 10⁻⁶–0.013)	Not available[a]
1,2-Dichloropropane	0.0013 (7.2 × 10⁻⁶–0.014)	Not available[a]
Styrene	0.0013 (8.1 × 10⁻⁶–0.013)	Not available[a]
1,2-Dichloroethane	0.0016 (9.4 × 10⁻⁶–0.016)	4.2x10⁻⁸ (2.5 × 10⁻¹⁰–4.1 × 10⁻⁷)
1,2,4-Trichlorobenzene	0.00091 (5.8 × 10⁻⁶–0.012)	Not available[a]
1,1,2-Trichloroethane	0.0013 (6.9 × 10⁻⁶–0.014)	2.1x10⁻⁸ (1.1 × 10⁻¹⁰–2.2 × 10⁻⁷)

Values are presented as: Mean (2.5th percentile – 97.5th percentile).

[a]"Not available" means that no Inhalation Unit Risk value was available for calculation of the excess lifetime cancer risk.

Hazard Quotients for Subchronic Inhalation Exposure

Exposure concentrations and hazard quotients for a subchronic inhalation exposure scenario, for the Stagnant Two-Film Model are presented in Table 4.

No hazard quotients exceeded 1. 1,2-dichloropropane had the highest mean hazard quotient, with the next highest being benzene. 1,1,1-trichloroethane had the lowest mean hazard quotient.

11 of the 12 VOCs can affect the nervous system (all but 1,2,4-trichlorobenzene). A summation of the available mean hazard quotients for the chemicals affecting the nervous system provides a hazard index of 0.089 (0.00073–1.0). While the mean hazard index did not exceed 1, the 97.5 percentile value did reach 1.0, indicating potential for adverse health impacts.

Excess Lifetime Cancer Risks

Exposure concentrations and excess lifetime cancer risk values for the Stagnant Two-Film Model are presented in Table 5.

The excess cancer risk did not reach the upper limit of the range of acceptable risk (10^{-4}–10^{-6}) for any of the individual VOCs for which they could be calculated. Of the 5 chemicals with cancer risk values, benzene had the highest mean exposure concentration and 1,1,2-trichloroethane had the lowest. 1,2-dichloroethane had the highest excess cancer risk and carbon tetrachloride had the lowest.

The cancer risk values were so low that all 5 values were summed to provide a total excess cancer risk of 1.04×10^{-7} ($1.4 \times 10^{-9} - 1.03 \times 10^{-6}$). These values do not exceed the general acceptable excess cancer risk range, even at the 97.5 percentile.

DISCUSSION

The individual concentration values of the 12 VOCs demonstrated no unacceptable increased risk of adverse health effects. No VOC demonstrated a hazard quotient greater than 1; hazard indices did not breach unity. The mean excess cancer risks due to these inhalation exposures appear to be no higher than about 4 in 100,000,000 (10^{-8}) for any individual chemical. None of the 97.5 percentile calculations exceeded the unacceptable risk ranges; the highest cancer risk was found at the 97.5 percentile for 1,2-dichloropropane (approximately 4 in 10,000,000 (10^{-7})). Combined excess cancer risks reached no higher than approximately 1.03 in 1,000,000 at the upper bound.

In McKenzie et al., 78 VOCs were evaluated in air emissions, significantly more than studied here; not all of which necessarily arose from flowback. Some of the highest contributors to the rise in adverse subchronic health effects, though, were also the same as those studied here, namely, aliphatic hydrocarbons and xylenes. Air concentrations and hazard quotients could not be compared to this study, as the study by McKenzie et al. presented data for median air concentrations and 95% upper confidence limits of mean concentrations (IOM and Adgate, 2012a and McKenzie et al., 2012).

Of the VOCs measurable in the Barnett Shale (Bunch et al., 2014), risk assessments showed hazard indices below the risk level of 1. Cancer risk ranges were found to be within the acceptable risk range of 10^{-4} to 10^{-6}. Only 6 VOCs were named as hydraulic fracturing-related (benzene, ethylbenzene, m/p-xylene, o-xylene, n-hexane, and toluene), 5 of which were covered in this study, but up to 105 VOCs were measured overall. Average air concentrations in Bunch et al.'s (2014) study were much lower than those calculated here.

This study involved calculated modeling, not air exposure measurements. Additionally, the contributions of other VOCs not measured here can affect the risk values. The relative contributions of the chemicals without inhalation unit risk values to the cancer risks of these VOC exposures are also unknown. Other limitations, such as the assumptions of constant weather conditions, the exposure period, and the EPA-provided standards for use in its RAGS Part F, present controls that likely vary with company and site practice (e.g. one pit or multiple pits), regulation, geography, geology, climate, and worker habits. An attempt was made, however, to conduct sensitivity analyses to consider other conditions: subchronic exposures of 9-months each year for 10 years, the presence of 2 pits, and two changes in wind speed were all modeled.

For multiple exposures in a 10-year period, the excess lifetime cancer risk increased approximately by a factor of 10 (data not shown). For example, benzene's excess lifetime cancer risk increased from 2.0×10^{-8} [$8.2 \times 10^{-11} - 2.2 \times 10^{-7}$] to 2.7×10^{-7} [$1.1 \times 10^{-9} - 2.9 \times 10^{-6}$]. Hazard quotients decreased slightly (e.g. 1,2-dichloropropane decreased from 0.047 [0.00025–0.5] to 0.035 [0.00019–0.37]), which was due to the incorporation of 3 months in each year during which no exposure occurred. No risk values exceeded the acceptable risk ranges.

The presence of 2 pits was evaluated by doubling the exposure concentrations. This resulted in an approximate doubling of all risk values (data not shown). For example, the hazard quotient for 1,2-dichloropropane increased from 0.047 [0.00025–0.5] to 0.093 [0.0005–0.99]. The individual increases in hazard quotients resulted in an increase in hazard index that created an exposure of concern at the 97.5 percentile (0.18 [0.0015–2.0]); however, no other unacceptable risk levels for hazard quotients or hazard index were breached at the 2.5 percentile or mean concentrations. None of the excess lifetime cancer risks exceeded the acceptable risk range.

With a decrease in wind speed to 1 mph, all risk values increased by a factor of 10, placing the hazard indices and some of the hazard quotients in the range of concern for adverse health effects; cancer risks still did not breach the range of acceptable risk. For example, benzene increased to 0.19 [0.00076 – 2.03]. The hazard index increased to 0.89 [0.0073 – 10.0]. A wind speed of 20 mph decreased all risk values by a factor of 2 (e.g. benzene decreased to 0.0093 [$3.8 \times 10^{-5} - 0.102$], and the hazard index decreased to 0.045 [0.00037–0.50]).

The data presented here appear to be more consistent with that presented by Bunch et al. (2014). All studies presented here are limited by variability in weather conditions, industry practice, study conditions, and well-specific practice.

This study does not eliminate concern for the additive effects of multiple small exposures to multiple chemicals (beyond the 12 studied here), nor does it eliminate the concern for the combined effects of flowback emissions with other sources of emissions at the well site.

CONCLUSIONS

Taken alone, our analysis of human health risk through air exposure to VOCs from flowback pits do not point to the need to prescribe alternative strategies of flowback containment or to recommend specific respiratory protections for workers. While it is reassuring to see that many of the calculations did not show an increased unacceptable risk, the limitations of the models preclude any definitive conclusions regarding worker risks. In particular, a sensitivity analysis indicates that low wind conditions may present concerns. In addition, since only 12 volatiles were studied in a simplified weather scenario, it would be premature to recommend for or against the use of additional protective measures to protect workers from inhaled exposures. These 12 VOCs were selected as the highest priority VOCs based on a previous study (Abualfaraj et al., 2014), so to the extent that the prioritization is correct, other VOCs would be expected to present lower risks. It is recommended, then, that further assessment occurs regarding the inhalation health effects of flowback volatiles as well as the individual and combined effects of these 12 VOCs and the remaining chemicals in flowback. This research should involve on-site air sampling to determine actual worker exposures, longitudinal studies of workers over time, on-site monitoring of worker health, and toxicological studies of this mixture of VOCs. It is also recommended that other exposure pathways (e.g. skin exposure) be evaluated further, since inhalation may not be the exposure pathway of most concern nor may it be the only exposure pathway.

The only instance in which risk values for the 12 VOCs exceeded standard risk levels for adverse health effects at the 97.5 percentile was under the assumption of low wind conditions and the presence of multiple pits. It is therefore recommended that further study of industry practice, for example, regarding pit size and number, as well as well site location, be performed to provide a more accurate picture of worker exposure scenarios. Further evaluation of additional VOCs and the 12 of interest here may provide more information regarding the risks depending upon various conditions of weather, industry, regulation, and location.

The limitations presented by this study require further research, including toxicologic and on-site studies. Flowback pits cannot definitively be declared safe based upon these results alone, but the use of additional respiratory protection measures is not an appropriate recommendation to make based only on the results found here.

ACKNOWLEDGMENTS

Mark Weir contributed to initial discussions on this topic and identified volatilization from flowback storage reservoirs as an issue of concern. Dr. Esther Chernak provided valuable advice, support, and writing assistance.

REFERENCES

1.   Abualfaraj, N., Gurian, P., Olson, M., 2014. Characterization of Marcellus shale flowback water. Environ. Eng. Sci. http://dx.doi.org/10.1089/ees.2014.0001.

2.   Agency for Toxic Substances and Diseases Registry (ATSDR), 2012. Toxic Substances Portal: Toxicological Profiles. Retrieved from. http://www.atsdr.cdc.gov/ toxprofiles/index.asp#X.

3.   Agency for Toxic Substances & Disease Registry (ATSDR), 2014. Priority List of Hazardous Substances: the Priority List of Hazardous Substances that Will Be the Subject of Toxicological Profiles. Retrieved from. http://www.atsdr.cdc.gov/ spl/.

4.   American Petroleum Institute (API), 2010. Water Management Associated with Hydraulic Fracturing: API Guidance Document HF2, first ed. Retrieved from. http://www.api.org/~/media/Files/Policy/Exploration/HF2_e1.pdf.

5.   American Petroleum Institute (API), 2011. Practices for Mitigating Surface Impacts Associated with Hydraulic Fracturing: API

Guidance Document HF3, first ed. Retrieved from. http://www. shalegas.energy.gov/resources/HF3_e7.pdf.

6.   Bloomdahl, R., 2014. Assessing Worker Exposure to Inhaled Volatile Organic Compounds from Marcellus Shale Flowback Pits (Unpublished master's thesis). Drexel University, Philadelphia, Pennsylvania.

7.   The Bousson Advisory Group, 2012. How Hydraulic Fracturing Works. Retrieved from. http://sites.allegheny.edu/ boussonadvisorygroup/how-hydraulicfracturing-works/.

8.   Brady, W.J., 2012. Hydraulic Fracturing Regulation in the United States: the Laissezfaire Approach of the Federal Government and Varying State Regulations. Retrieved from. http://www.law.du.edu/ documents/faculty-highlights/Intersol- 2012-HydroFracking.pdf.

9.   Bunch, A.G., Perry, C.S., Abraham, L., Wikoff, D.S., Tachovsky, J.A., Hixon, J.G., et al., 2014. Evaluation of impact of shale gas operations in the Barnett Shale region on volatile organic compounds in air and potential human health risks. Sci. Total Environ. 468e469, 832e842. http://dx.doi.org/10.1016/ j.scitotenv.2013.08.080.

10. Caryl-Sue (Producer), 2013, March. National Geographic Magazine. How Hydraulic Fracturing Works [video]. Retrieved from. http://education.nationalgeographic. Com/education/ media/how-hydraulic-fracturing-works/? ar_a¼5.

11. Chapter 78, 2001. Title 25, the Pennsylvania Code, 25 x Pa. Code 78.56. Retrieved from. http://www.pacode.com/secure/data/025/ chapter78/chap78toc.html.

12. CHEMnetBASE, 2010. Combined Chemical Dictionary. Retrieved from. http://www. chemnetbase.com.ezproxy2.library.drexel. edu/.

13. Colborn, T., Kwiatkowski, C., Schultz, K., Bachran, M., 2011. Natural gas operations from a public health perspective. Hum. Ecol. Risk Assess. Int. J. 17 (5), 1039e1056. http://dx.doi.org/10. 1080/10807039.2011.605662.

14. Ground Water Protection Council (GWPC), & Interstate Oil & Gas Compact Commission (IOGCC), 2014. FracFocus: Chemical Disclosure Registry. Retrieved from. http://fracfocus.org/.

15. Institute of Medicine of the National Academies (IOM) (producer), Adgate, J. (writer), 2012a. Air Pollution Exposure and Risk Near

Unconventional Natural Gas Drill Sites: Example from Garfield County, Colorado [video]. Retrieved from. http://www.iom.edu/ Activities/Environment/EnvironmentalHealthRT/2012- APR-30/ Day-1/Session-5/3-Adgate.aspx.

16. Institute of Medicine of the National Academies (IOM) (producer), Esswein, E. (writer), Breitenstein, M. (writer), Snawder, J. (writer), 2012b. NIOSH Field Effort to Assess Chemical Exposures in Oil and Gas Workers: Health Hazards in Hydraulic Fracturing [video]. Retrieved from. http://www.iom.edu/Activities/ Environment/ EnvironmentalHealthRT/2012-APR-30/Day-1/Session-3/1-Esswein.aspx

17. Lewis Sr., R.J., 2007. Hydraulic fracturing. In: Hawley's Condensed Chemical Dictionary, fifteenth ed. John Wiley & Sons, Inc, Hoboken, NJ, p. 661.

18. McKenzie, L.M., Witter, R.Z., Newman, L.S., Adgate, J.L., 2012. Human health risk assessment of air emissions from development of unconventional natural gas resources. Sci. Total Environ. 424, 79e87. http://dx.doi.org/10.1016/ j.scitotenv.2012.02.018.

19. Navarro, G.L., 2011. Earth Sciences in the 21st Century: Marcellus Shale and Shale Gas: Facts and Considerations. Nova Science Publishers, Inc, New York, NY.

20. Occupational Safety and Health Administration (OSHA), 29 Jun 2004. Standards Interpretation: 01/17/2002 e Exemption under the Benzene Standard for Oil and Gas Drilling Operations. Retrieved from. https://www.osha.gov/pls/ oshaweb/owadisp. show_document?p_table¼INTERPRETATIONS&p_id¼24259.

21. The Penn State Marcellus Center for Outreach and Research (MCOR), 2014. Resources: FAQs. Retrieved from. http://www. marcellus.psu.edu/resources/faq. php.

22. Schramm, E., 2011. The Institute for Energy & Environmental Research for Northeastern Pennsylvania: the Marcellus Shale Clearinghouse: what Is Flowback, and How Does it Differ from Produced Water?. Retrieved from. http://energy. wilkes.edu/ pages/205.asp.

23. Schwarzenbach, R.P., Gschwend, P.M., Imboden, D.M., 1993. Environmental Organic Chemistry. John Wiley & Sons, Inc, New York, NY.

24. Suchy, D.R., Newell, K.D., May 15, 2012. Kansas Geological Survey, Public Information Circular (PIC) 32: Hydraulic Fracturing of Oil and Gas Wells in Kansas. Retrieved from. http://www.kgs. ku.edu/Publications/PIC/pic32.html.

25. Title 40: Part 300 e National Oil and Hazardous Substances Pollution Contingency Plan. 59 Fed. Reg. 47, 416 (September 15, 1994). Retrieved from www.ecfr.gov/.

26. United States Department of Labor, Occupational Safety & Health Administration (OSHA), 2014a. Chemical Sampling Information. Retrieved from. https://www. osha.gov/dts/chemicalsampling/ toc/toc_chemsamp.html.

27. United States Department of Labor, Occupational Safety & Health Administration (OSHA), 2014b. OSHA Occupational Chemical Database: 1,2-Dichloropropane. Retrieved from. https://www. osha.gov/chemicaldata/chemResult.html? recNo¼677.

28. United States Environmental Protection Agency (EPA), 1998. AP42, Fifth Edition: Compilation of Air Pollutant Emission Factors, Volume I: Stationary Point and Area Sources: Chapter 4.3: Wastewater Collection, Treatment, and Storage. Retrieved fromhttp://www.epa.gov/ttnchie1/ap42/ch04/final/c4s03.pdf.

29. United States Environmental Protection Agency (EPA), Oct. 2006. List of Lists: Consolidated List of Chemicals Subject to the Emergency Planning and Community Right-to-know Act (EPCRA) and Section 112(r) of the Clean Air Act. Retrieved from. http:// www.epa.gov/oem/docs/chem/title3_Oct_2006.pdf.

30. United States Environmental Protection Agency (EPA), 2009. Risk Assessment Guidance for Superfund: Volume I: Human Health Evaluation Manual (Part F, Supplemental Guidance for Inhalation Risk Assessment) (EPA-540-R-070e002). Retrieved from. http:// www.epa.gov/swerrims/riskassessment/ragsf/index.htm.

31. United States Environmental Protection Agency (EPA), 2011a. Exposure Factors Handbook: 2011 Edition. Retrieved from. http:// www.epa.gov/ncea/efh/pdfs/ efh-complete.pdf.

32. United States Environmental Protection Agency (EPA), 2011b. Mid-atlantic Oil and Gas Extraction. Retrieved from. http://www. epa.gov/region03/marcellus_shale/ index.html.

33. United States Environmental Protection Agency (EPA), 2011c. Technology Transfer Network Air Toxics: 2005 National-scale

Air Toxics Assessment: Glossary of Terms. Retrieved from. http://www.epa.gov/ttn/atw/natamain/gloss1.html.

34.   United States Environmental Protection Agency (EPA), 2013a. EPA On-line Tools for Site Assessment Calculation. Retrieved from. http://www.epa.gov/athens/ learn2model/part-two/onsite/ estdiffusion-ext.html.

35.   United States Environmental Protection Agency (EPA), 2013b. Natural Gas ExtractiondHydraulic Fracturing. Retrieved from. http://www2.epa.gov/ hydraulicfracturing.

36.   United States Environmental Protection Agency (EPA), 2014a. Integrated Risk Information System (IRIS). Retrieved from. www.epa.gov/IRIS/.

37.   United States Environmental Protections Agency (EPA), 2014b. Provisional Peer Reviewed Toxicity Values for Superfund (PPRTV): PPRTV Assessments Electronic Library. Retrieved from. http://hhpprtv.ornl.gov/.

38.   United States National Library of Medicine (NLOM), 2011. TOXNET: Toxicology Data Network: Hazardous Substances Data Bank (HSDB). Retrieved from. http:// toxnet.nlm.nih.gov/cgi-bin/ sis/htmlgen?HSDB.

Chapter 5

A Review on the Fuel Gas Cleaning Technologies in Gasification Process

Prabhansu[a], Malay Kr. Karmakar[b], Prakash Chandra[a], and Pradip Kr. Chatterjee[b]

[a]Mechanical Engineering Department, National Institute of Technology, Patna 800005, India

[b]Thermal Engineering, CSIR – Central Mechanical Engineering Research Institute, Durgapur 713209, India

ABSTRACT

The product gas produced from gasification of solid fuel contains various impurities such as particulates, toxin gases, tar, vapours of heavy metals, etc. Presence of tar is a major issue which requires to be addressed before the use of gas product in the downstream process. Tar causes problems in the process equipment like flow channels, power generating units, etc. Generally gasification technology is adapted for the utilization of low grade coal, municipal solid waste, agro-waste, bio-waste, etc., which generates toxic and emits various hazardous compounds of chlorine, sulphur, nitrogen and heavy metals like Mn, Cd and Hg. Various alkali metals like Na, K, etc., generated through the gasification of wastes also create problem in the downstream processes when condensed at low temperature. The key challenge to commercializing gasification technology is to generate a clean fuel gas which meets the global emission standards. This paper provides a comprehensive overview of the fuel gas cleaning methods those are used to remove the contaminants and gas impurities generated from various types of reactor for gasification of coal or biomass.

INTRODUCTION

India is the sixth largest energy consumer in the world, accounting for 3.4% of the global energy and the major part of the energy consumption is contributed by coal. The country has 298.94 billion tonnes of coal and 43.22 billion tonnes of lignite reserves as on 2013 [1]. Therefore, the use of coal is bound to increase or at least remain steady for years to come and the demand for electricity in India is expected to exceed 950,000 MW by 2030. Another severe issue is that pollution level at several major cities in India is on rise. The increasing demand of electricity and rise of pollution level should simultaneously be considered as both are the wheels of same vehicle, i.e., development. Therefore, the clean fuel technology is the need of hour [1], [2] and [3].

In the current scenario, one solution could be to introduce gasification technology instead of the conventional direct burning of coal in power plants which creates a lot of health hazards [4]. The gasification process converts organic materials or fossil fuel based carbonaceous compound into carbon monoxide, hydrogen and carbon dioxide, which is achieved by reacting the solid fuels at high temperatures with a controlled amount of oxygen and/or steam. Gasification of fossil fuels is currently being used on industrial scales to generate electricity in some countries, but gradually biomass and low grade coal gasification is bound to gain momentum in the coming years. Fuel gas could also be produced at the site of coal mines through gasification and may be sent back to the power plants or household as a clean fuel. Fuel gas production at site is also beneficial to the mankind as there is no toxic emission unlike conventional power plants. Other advantage of gasification process is that it does not contain the toxic compounds like dioxins and furans unlike the direct combustion process [5] and [6].

Despite the numerous advantages of gasification process, the technology is still in the developing stage due to some challenges. Impurities such as tars, particulate matters, NH_3, H_2S, HCl and SO_2, which are unavoidably produced during gasification and generally sustained in the producer gas, cause severe problems in downstream applications [7], [8], [9], [10] and [11]. These contaminants must be removed before the gas is being used for internal combustion engine, fuel-cell and for secondary conversion into liquid fuels or chemicals by Fischer–Tropsch synthesis [12], [13], [14] and [15].

The present work focuses on the study of removal processes of the particulate matters, tar, sulphur compounds, alkali and heavy metals, nitrogen and chlorine compounds from raw fuel gas in gasification process. A comprehensive literature survey of the above subject has been conducted.

Impurities in Fuel Gas and Environmental Pollution

The fuel gas from gasification processes contains various amounts of impurities and particulate matters originating from the solid fuel and particle attritions from the bed. There are certain elements inherently present in solid fuel which creates pollution to the environment when it is burnt or gasified. Prominent among them are sulphur, chlorine, alkali metals (sodium and potassium) and several other heavy metals like Ba, Zn, Ni, Cu, Fe, Pb, Mn, Mg, and Cd. These toxic elements are of concern today as they are creating a lot of health hazards, both on human body as well as on livestock. Although the toxic effects of alkali metals have negligible effects on living being, still it creates a lot of problem during downstream applications. Alkali metals principally potassium and sodium and alkaline earth metals like calcium, etc., are present in biomass and coal. These elements have the tendency to vapourize at temperature in thermal conversion facility where they react with silica, sulphur and chlorine. Downstream heat recovery reduces product temperature. This results in condensation of inorganic compounds that cause deposits, fouling and corrosion [16].

Table 1 shows the major pollutants coming out from the emission of coal burning/gasification and the problems created on human body. Advanced technologies today partially offer solutions to some of these problems. For raw fuel gas cleaning there are basically three techniques, i.e., hot gas cleaning which operates above 300 °C, cold gas cleaning which operates below 100 °C, while warm gas cleaning operates in between the two extremes [17] and [18]. Especially, the integration of these systems in larger scales has to be surveyed more carefully. For a successful implementation of gasification in commercial fuel gas production, the effluent gas must conform to allowable limits regarding particulate and other impurities.

Table 1: Effect of toxic elements on human beings

Toxic elements	Effect on human beings
Dioxin/furans	Cloracne, reproductive, developmental problems, damages the immune system, interfere with hormones and also causes cancer [19].
Sulphur oxides	Asthma, chronic bronchitis, air ways inflammation, eye irritation, psychic alteration, heart failure [20].
Nitrogen oxides	Damages cell membranes in the lung tissue, causes constriction of the lung way passage, nasal irritation and pulmonary discomfort is common [20].
Arsenic	Induces reactive oxygen species (ROS) and oxidative stress. Binds to thiols, alters signal cascade, imbalance in antioxidant levels. Triggers apoptosis & cell death, nausea, decrease in production of red and white blood cells, sensation of pin and needles in hand & feet [21].
Chromium	Nose ulcers, skin ulcer, asthma, shortness of breath or wheezing, allergic reactions leading to severe redness and swelling of skin, long term exposure leads to damage of kidney, circulating and nerve tissues as well as skin irritation [22] and [23].
Copper	Gastrointestinal: metallic taste, nausea, vomiting, gastrointestinal bleeding, Renal: haematuria, oliguria, elevated urea, creatinine & acute tubular necrosis [22] and [23].

Cadmium	Gastrointestinal symptoms: nausea, vomiting, abdominal pain, diarrhoea, salivation, tenesmus, haemorrhagic gastroenteritis, may result in pulmonary fibrosis, headache, cadmium fume pneumonitis, hepatic necrosis, Renal necrosis, cardiomyopathy, chills, weakness & dizziness [22] and [23].
Lead	Damages the brain and kidney, in pregnant women, high level of exposure may cause miscarriage, high level exposure in men can damage the organs responsible for sperm production [24].
Mercury	Exposure to high level can permanently damage the brain, kidney and developing foetuses, effects on brain functioning may result in irritability, shyness, tremors, changes in vision or hearing and memory problems[21].

Removal of Fuel Gas Impurities

Particulate Matters Removal

Particulate matter in gasifiers ranges from less than 1 μm to more than 100 μm and it varies on the basis of composition and type of feedstock [25]. Classification of particulate matters is done on the basis of aerodynamic diameter. Residual carbon and inorganic compounds constitute the bulk of particulate matters. Alkali metals (Na and K), alkaline earth metals (Ca), silica (SiO_2) and other metals like Fe and Mg compounds constitute the inorganic materials [26] and [27]. Minor constituents in particle matters include arsenic, selenium, antimony, zinc and lead [28]. The critical issues with particulate matters involve fouling, corrosion and erosion of downstream components like gas turbine blades, etc., [29], [30] and [31]. There are three different techniques used for its removal.

Hot Gas Particulate Matters Removal

The removal techniques of particulate matters from fuel gas are mainly based on three categories, i.e., inertial separation, barrier filtration and electrostatic separation [32].

Inertial Separation

It operates using mass and acceleration principles for the separation of heavier solids from lighter gases. Three significant devices of this category are impact separators, dust agglomerators and cyclone, but the most trusted one is cyclone among them [28]. Cyclones utilize centripetal force to create vortex for particulate removal which can comfortably operate at 1000 °C and is one of the most widely used techniques for particulate separation.

The design of cyclone is based on the characteristics of the particle and the gas stream. Different approaches have been developed in the past [33]. A 'cut point' is made so that the particle may obtain a balance between centrifugal and drag force. The cut point is the particle size at this point and denoted by X50 or d50 which means that it has the removal efficiency of 50% [25]. Although cyclones are mature technology, the separation efficiency of reverse flow cyclone using partial recirculation is more than the conventional Stairmand high efficiency designs. Particulate removal surpasses 99.6% in cyclones when compared with low temperature high efficient devices such as venturi and pulse jet bag filters [34]. As per Lee et al. [35], a high-temperature and high-pressure dust collector has been tested at 800 °C and 3 atm and the overall collection efficiency is found to be over 99.999%.

Filters

When a gas stream passes around granules or through a porous monolithic solid, the filters are known as barrier filters. During filtration, particulate matters are separated in four steps which are diffusion, inertial impaction, gravitational settling and particle collection. It occurs as a result of random collisions with the filter media as they deviate from the gas streamlines [31].

Ceramic or metallic materials are the common raw materials for the construction of rigid filters. They have the capability to remove 99.99% of particulate matter of size smaller than 100 μm with operating temperature above 400 °C [36]. There are candle filters which are made up of clay-bonded silicon carbide (SiC) as well as monolithic and composite ceramics and can sustain high temperature [37]. Metals may also be introduced to prevent ceramics from damage in hostile conditions and also to provide catalytic activity [38]. Ceramic filters are fragile, which led to the development of sintered metal barrier filters where the operational temperature may be raised to 1000 °C. The concentrations of particulate matters could be as low as 10 mg/m³ with filtration efficiencies approaching to almost 100% [39]. For the construction of metallic filters, metallic powders (iron aluminide) are heated within a mould to a temperature, where the material begins to fuse together [40].

Fixed or moving granular bed filters may avoid the limitations of the metallic and ceramic filter elements. Granular materials such as low-silica lapilli (volcanic rock), limestone or sand are placed inside a vessel through which particulate laden gas may be passed. Sintered bauxite (alumina oxide) with particle diameters typically in the order of several hundred micrometers could be one of the choices [41]. Factor like flow rate of the granular media also affects the efficiency of the moving bed filters. Pressure drop can also be altered by thickness of dust cake [42]. Several other factors which affect the efficiency are the shape of solids and loading of the particulates in fuel gas, filter's gas velocity which should be kept below the minimum fluidization velocity of the granules [43].

Different granular media like alumina and mullite are used for enhancing removal with electrostatic forces. It has been observed that using this, efficiencies are as high as 99% for 4 µm particles and 93% for 0.3 µm size particles at temperature up to 840 °C [44]. Standleg moving granular bed filter system (SMGBF) has shown efficiency of above 99.9% even at 870 °C [45]. Moving granular bed filter is a promising future for high temperature and robust operation with minimal maintenance [46]. Catalytic material can also be incorporated into filter material for simultaneous removal of particulate matter and tars. A catalytic filter created by adding nickel (Ni) and magnesium oxide (MgO) to the pores of an -alumina (Al_2O_3) candle filter have shown to improve tar reduction capabilities [47]. In another experiment, a silicon carbide filter was loaded with $MgO–Al_2O_3$ supported Ni catalyst was placed in the gasifier for in-situ reduction of tar. It was observed that 58% tar and 28% methane were converted to yield improved hydrogen content with overall increase in gas yield by 15% [48].

Electrostatic Separator

In electrostatic separations, electric properties are exploited to remove particles from gas flows. Due to strong electric field, particles become charged and according to the difference in dielectric properties they are removed. Electrostatic forces acting on particles (<30 μm) are 100 times stronger than gravitational force. As a result of that electrostatic precipitators are very efficient in particulate matter removal [49]. ESPs have conventionally been used for fly-ash removal in coal fired power plants at temperature up to 200 °C[50]. The electrostatic precipitator (ESP) can also be used in gasification plants between 300 and 450 °C for oil vapour separation [51]. However, use of ESPs is very limited for high temperature application as the temperature largely affects the viscosity, density and resistivity of the gas and thereby causes to deteriorate the performance of ESPs.

Several other non thermal plasma techniques related to electrostatic precipitators have been developed for gas cleaning which includes pulsed corona, dielectric barrier discharges, DC corona discharges, microwave and RF plasma. In these systems, high energy pulse creates high energy electrons, which then generates other electrons and ions. Its operation is similar to that of an ESP [52].

Warm Gas Particulate Matters Removal

Three techniques are commonly used for removal of particulates in warm gas condition which are namely gas cyclones, ESP and filtration devices. Several types of filtration devices exist including fabric filters, rigid filters and both fixed and moving bed granular filters. Fabric filters effectively remove particulate matters larger than 1 μm and concentrations less than 1 μg/m^3[31]. Due to construction material of these filters, operation temperature is limited to 250 °C which classifies them as warm gas cleanup [31] and [53]. The fabric filters have got the limitation that it can be used only between 90 and 250 °C in order to limit its exposure to liquids in the gas stream [31]. Fig. 1 shows the particle separation efficiencies of some gas cleaning systems.

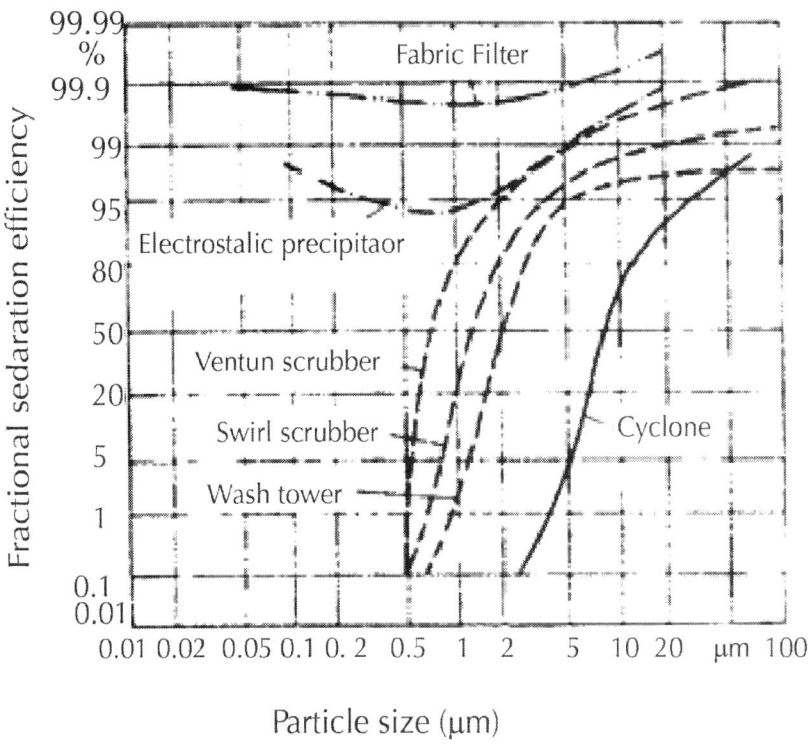

Figure 1: Particle separation efficiencies of some gas cleaning systems [54].

Cold Gas Particulate Matters Removal

Particulate matters may also be removed by wet scrubbing technology at ambient temperature using water. Cold gas scrubbing technology is categorized as wet dynamic scrubbers, spray scrubbers, cyclonic spray scrubbers, impactor scrubbers, venturi scrubbers and electrostatic scrubbers. It becomes more effective as particle reaches the size of 3 μm [55]. Wet ESPs, however, has several disadvantages due to complexity, residual waste stream and other cold gas cleaning problems. Table 2 shows the operating conditions and efficiency of particulate removal equipments at different operating temperatures.

Table 2: Operating conditions and efficiency of particulate removal equipments [56]

Dust separator	Temperature range (°C)	De-dusting efficiency	Pressure drop (kPa)
Cyclone	100–900	Dust >5 µm, 80%	<10
Fabric bag filters	60–250	Dust >0.3 µm, 99–99.8%	1–2.5
Wet scrubbers (venturi)	20–100	Dust >0.1–1 µm, 85–95, otherwise 90–99%	5–20
Fibrous ceramic filters	200–800	Dust >0.1 µm, 99.5–99.99%	1–5
Metallic foam filters	200–800	Dust >1 µm, 99–99.5%	<1
Granular bed filters	200–800	Highly depends on regime and surface cake filtration.	<10

Tar Cleaning

Tar may be described as a complex mixture of condensable hydrocarbons, which are largely aromatic. It includes single to multiple ring aromatic compounds along with other hydrocarbons and complex polycyclic aromatic hydrocarbons [57]. European commission (DG XVII) and US DoE have defined tar as hydrocarbons with molecular weight higher than benzene [58]. Tars may be classified into four classes, (a) primary products, derived from cellulose, hemicelluloses and lignin, (b) secondary products such as phenolics and olefins, (c) alkyl tertiary products which are basically methyl derivative of aromatic series, and (d) condensed tertiary poly-aromatic hydrocarbons (PAH) without substituent [59] and [60]. Tar concentration in fuel gas basically depends on the type of gasifier in use. In an air blown circulating fluidized bed (CFB) biomass gasifier, tar content is about 10 g/m^3. Tar content varies from about 0.5 to 100 g/m^3 in other type of gasifiers [61].

The tar removal process may be classified into two categories – it can be primary (inside the gasifier) or secondary (outside the gasifier) depending on the locations. Generally secondary method is used as it is most suitable [62]. Tar may be present in two forms, liquid droplet when the exit temperature is low (fixed bed updraft) and vapour when the exit temperature (downdraft and fluidized bed) is high [63]. For fuel gas cleaning again there are three basic classifications based on fuel gas temperatures – hot gas cleaning, warm gas cleaning employs the oil based gas washing (OLGA) technique and wet gas cleaning technology includes basically wet scrubbers [64].

Hot Gas Tar Cleaning

Hot gas tar cleaning technology includes four basic approaches, i.e., thermal cracking, catalytic cracking, non-thermal plasmas and physical separation.

Thermal cracking uses high temperature (1100–1300 °C) to decompose large organic compounds into small non-condensable gases [65]. For example, naphthalene may be reduced by more than 80% in about 1 s at 1150 °C, but it takes 5 s at 1075 °C [66] and [60]. In another experiment it was shown that only 0.5 s is required for tar to be reduced effectively at 1250 °C [67]. The disadvantage of thermal cracking is that as it approaches to downstream of gasifier, it may increase the soot production, which may also increase the particulate load on the processing equipment [60] and [68].

Tar decomposition occurs through either of the four processes.

Cracking: $pn_c HX \rightarrow qCmHy + rH_2$

$$\text{Reforming of steam}: C_nH_X + nH_2O \rightarrow \left(n + \frac{x}{2}\right)H_2 + nCO$$

$$\text{Dry reforming}: C_nH_X + nCO_2 \rightarrow \left(\frac{x}{2}\right)H_2 + 2nCO$$

$$\text{Carbon formation}: C_nH_X \rightarrow nC + \left(\frac{X}{2}\right)H_2$$

where C_nH_X represents tar, C_mH_y represents hydrocarbons with smaller carbon number than C_nH_X [69].

Catalytic cracking occurs at temperature lower than that of thermal cracking process. On the other hand, the catalyst activity gets reduced with time due to poisoning, fragmentation or carbon deposition [70]. These catalysts may be classified in several ways – it may be acidic, basic, iron based or nickel based catalyst [71]. Calcined dolomite is frequently used as tar cracking catalyst. This is created by heating dolomite to release CO_2 and have shown up to 95% tar conversion [72]. If iron and magnesium are added to the silicate mineral, it combines to form olivine $((Mg, Fe)SiO_4)$ with enhanced catalytic property. It has been found that olivine has greater attrition resistance as in-situ catalyst than dolomite, when applied to the gasification environment [73]. Olivine-supported nickel silicate, prepared by thermal impregnation, has been shown to be a potential tar cracking catalyst and exhibits good catalytic activity when used for dry reforming, water gas shift, reverse water gas shift, methanation reactions and steam reforming reaction[74]. Iron based catalysts often exist as oxides, representing approximately 35–70% mineral composition. They have many disadvantages which paved way for Ni based catalysts [72] and [75]. Table 3 describes the properties of commercially available nickel based catalysts. Nickel catalysts are quite effective in decomposing tar and their activities are approximately 10 times higher than that of dolomite [72] and [76]. 96–98% naphthalene removal efficiency and 41–79% benzene removal efficiency were revealed while using a co-precipitated catalytic filter disc at gas velocity of 2.5 cm/s and 100 ppm of H_2S at 900 °C [77].Fig. 2 shows the tar conversion with different mineral catalysts.

Table 3: Properties of commercially available nickel based catalysts [76]

Catalyst	NiO content (wt%)	Support	Surface area (m2/g)	Reactor temperature (°C)	Initial tar conversion (%)
Heavy hydrocarbons reforming					

BASF G1-25/1	25	Cao–Al2O3–SiO2–K2O	16.4	785–850	98–99
BASF G1-50	20	MgO–CaO–Al2O3–SiO2–K2O	19.9	660–800	89–99
Haldor Topsoe R-67	15	MgAl2O4–SiO2–K2O	17	780–840	95–100
ICI Katalco 46-1	22	MgO–CaO–Al2O3–SiO2–K2O	16.2	700–875	73–100
Sud Chemie C11-NK	20–25	MgO–CaO–SiO2–Al2O3	8.8	600–900	–
Light hydrocarbons reforming					
BASF G1-255	12–15	–	2.5	785	88–97
BASF V1693	10.2	–	–	850–900	–
Haldor Topsoe RKS-1	15	MgAl2O4–SiO2–K2O	6.8	785–800	92
Haldor Topsoe R-67-7H	16–18	MgAl2O4	12–20	690–780	99
ICI Katalco 57-3	12	CaO–Al2O3–SiO2	2.9	–	69
Sud Chemie G-90LDP	14	CaO–Al2O3	–	850–900	99
Sud Chemie G-90EW	14	CaO–Al2O3	–	850–900	99
Sud Chemie G-90B	14	CaO–Al2O3	5–7	650–900	99
United catalyst C11-9-061	10–15	Al2O3	2.9	725–800	–

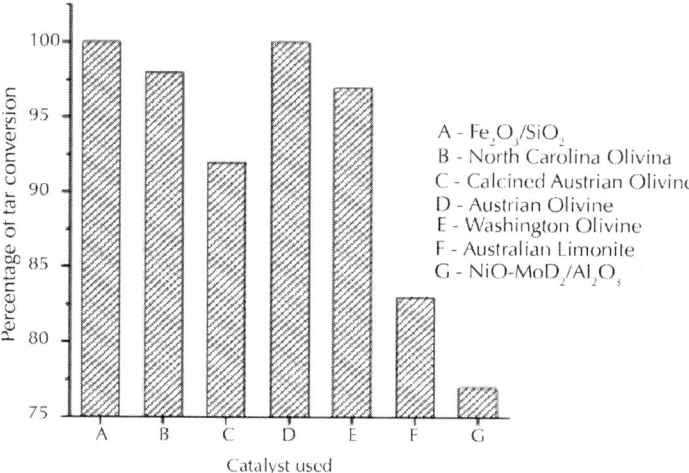

Figure 2: Tar conversion with different mineral catalysts [78].

In plasmas, presence of free radicals, ions and other excited molecules create a reactive environment that decomposes tar molecules. Two processes are followed for the generation of plasma, either through elevation of temperature much higher than that of the requirement in the gasifier (thermal plasma) or collision between high energy electrons and molecules (non-thermal plasma) [52]. Different types of non-thermal plasmas include pulsed corona, dielectric barrier discharges, DC corona discharges, RF plasma and microwave plasma. While these technologies have been effective, the cost, energy demand, device lifetime and operational complexities have limited their application [71]. Pulsed corona plasma has proved to be the most promising one that removes considerable tar at about 400 °C [79].

Partial condensation may be exploited for tar reduction in physical separation devices. The ideal temperature is around 450 °C. The tar condenses to form aerosols which have a larger mass and behave like the particulate matters. Hence, they can be separated by ESP and inertial separation devices [80].

By using rotating particle separator (RPS) and fabric filter, it was reported that 30–70% and 0–50% tar can be removed respectively [81]. In an experiment, hot gas ceramic filtration was studied where 77–97.9% tar reduction was observed while using quartz and 75.6–94% tar removal was found using glass fibre filter[82]. Fixed bed adsorbers like lignite coke or activated carbon precipitates tar from the fuel gas, using adsorption technique of high boiling tar compounds [83]. It was reported that 50–97% tar was reduced by using sand bed filter [81]. Barrier filters are not suitable for tar removal as tar gets deposited in the filter. This cannot be easily cleaned and may also lead to eventual plugging [82]. One of the recently developed techniques is catalytic filter [85]. This is a single step removal process of both particulate and tar. Nickel based candle filtration is found to be very effective at elevated temperature, i.e., above 850 °C [86].

Warm Gas Tar Cleaning

OLGA is actually a Dutch acronym for oil based gas washer and operates between 60 and 450 °C. It removes but reuses valuable tar components. OLGA technique uses oil rather than water to scrub tar. The patent holders for the OLGA techniques are the energy research centre of the Netherlands (ECN). Units are now commercially mature and have been successfully demonstrated at several facilities [76] and [87].Fig. 3 shows the schematic diagram of OLGA process.

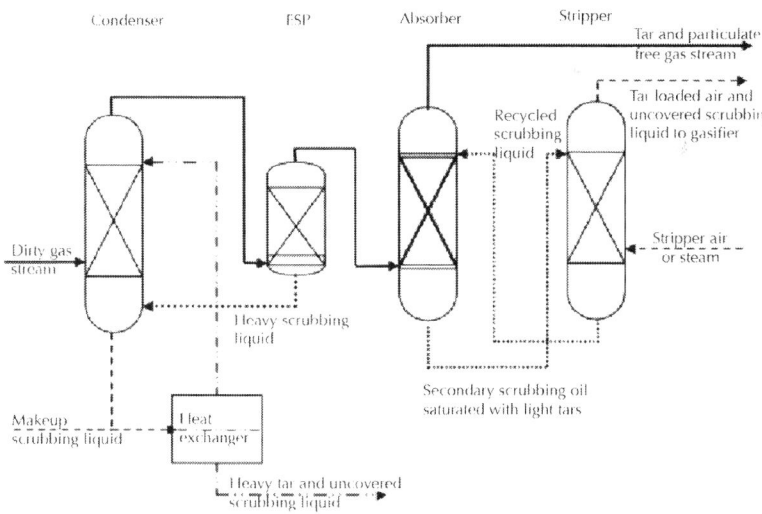

Figure 3: Simplified OLGA process diagram for tar removal [17], [87] and [88].

Cold Gas Tar Cleaning

It has been reported that the removal efficiency of a wet ESP can reach about 40–70% tar and more than 99% dust in an updraft gasifier at Harboore, a downdraft gasifier at Wiener Neustadt and CFB gasifier at ECN [89]. The ESP has been found to clean gas coming out from rotary kiln, saw mill, alkali by pass, clinker cooler, cement and coal mill [90]. Venturi scrubber has shown to be very efficient in removing tar and particle matters. It was also found that tar separation efficiency ranges from 50–90% in a venturi scrubber when it was used to clean the fuel gas from a countercurrent rice husk gasifier [91]. In Table 4, the tar and particulate matters removal efficiencies are shown for different methods.

Table 4: Tar and particulate removal efficiency of certain mechanism/methods [84]

Methods	Particulate removal (%)	Tar removal (%)
Sand bed filter	70–99	50–97
Wash tower	60–98	10–25
Venture scrubber	–	50–90
Wet electrostatic precipitator	>99	0–60
Fabric filter	70–95	0–50
Rotational particle separator	85–90	30–70
Fixed bed tar adsorber	–	50

Sulphur Compounds Removal

The main contaminants of sulphur in gasification based fuel gas are in the forms of hydrogen sulphide (H_2S), carbonyl sulphide (COS) and sulphur dioxide (SO_2). The tendency of sulphur based compounds is to corrode metal surfaces [92]. Even a small amount of sulphur is sufficient to poison catalysts which are used to clean or upgrade the fuel gas [93]. The concentration of H_2S in raw fuel gas depends on the feed stocks and varies between 0.1 mL/L to more than 30 mL/L of fuel gas [80] and [94]. Biomass contains less sulphur (0.1–0.5 g/kg) as compared to coal (50 g/kg) derived fuel gas [95].

Depending on the various fuel gas applications, H_2S has been made a prime focus for hot sulphur removal. Some applications of fuel gas require very low level of sulphur as low as pico litres per litre of fuel gas to avoid equipment failures. During burning of fuel gas or producer gas, sulphur contaminants are oxidized to sulphur dioxide. Historically sulphur removal at high temperatures is performed by scrubbing the SO_2 products emitted from a combustion process.

Hot Gas Sulphur Compounds Removal

Most of the hot gas cleaning technologies utilizes adsorption technique for the removal of sulphur based compounds. In this process, gaseous species are made to combine physically or chemically with the adsorbent in solid state. Physical adsorption involves weak van der Waal's intermolecular dipole interactions due to polarizations within molecules and are very weak to allow relatively easy desorption. Chemical adsorption involves covalent bonding of adsorbate molecules onto the surface of adsorbent while this process may be too strong for easy desorption of contaminants and occurs only where the sorbent surface is available for reaction. A sulphur adsorption process involves three essential steps, i.e., reduction, sulphidation and regeneration of adsorbent [96]. In the preparation step, the solid sorbent is first reduced for chemical adsorption with the sulphur species. In the sulphidation reaction, metal oxides (such as Zn or Fe) combine with sulphur to form ZnS or FeS. A reversible process is then followed with regeneration to recover original metal oxide and enriched sulphur dioxide gas stream. If the commercial installation is large enough, then it is directed to sulphur recovery where sulphuric acid or elemental sulphur is recovered.

After extensive research, a list of potential desulphurization metals has been obtained which are Zn, Fe, Cu, Mn, Mo, Co and V [96]. Some mixed oxides, such as Mn with V and Cu oxides, have shown good desulphurization at temperatures above 600 °C [97]. Zinc and Copper oxides, abundantly available in nature, are being used extensively for first stage sulphur removal and have the potential for removal of above

99% sulphur. ZnO in particular was originally developed for one-time use but it is now being used as a renewable sorbent [98]. Other combinations with CuO include Fe_2O_3 and Al_2O_3. A ZnO sorbent is very effective in removing H_2S even up to 10 µL/L, but begins to volatilize as average operational temperature reaches 600–650 °C [96]. Improvements in ZnO/CuO sorbents have led to its commercialization. Companies such as Sud-Chemie and ConocoPhillips are producing sorbents varying in composition from pure ZnO to mixtures of ZnO/ CuO/Al_2O_3 [99]. Both zinc and iron have an affinity for sulphur to provide zinc ferrites ($ZnFe_2O_4$) which has a high sulphur loading capacity of more than 300 g/kg of fresh catalyst. Iron based materials, however, have a drawback when they suffer from carbon deposition and further become worse with increased H_2O content during sulphidation reaction [71] and [100]. In another study [101], molten carbonate is shown to act as absorbent during desulphurization and dechlorination and as a thermal catalyst for tar cracking. Within the temperature range of 800–1000 °C, it is discovered that the removal of H_2S depends on the concentration of CO_2 in the fuel gas. When it is in trace amount, desulphurization using molten carbonate is inadequate. However, when carbon elements like char and tar were continuously supplied, H_2S removal can be very high.

Organic sulphur (COS) can be converted to H_2S using CoMo (Cobalt–Molybdenum) and nickel-based catalysts. The conversion efficiency of COS is shown in Fig. 4. This figure shows that the removal of COS on CoMo catalysts decreases greatly with time whereas the nickel based catalyst remains stable and active over the time.

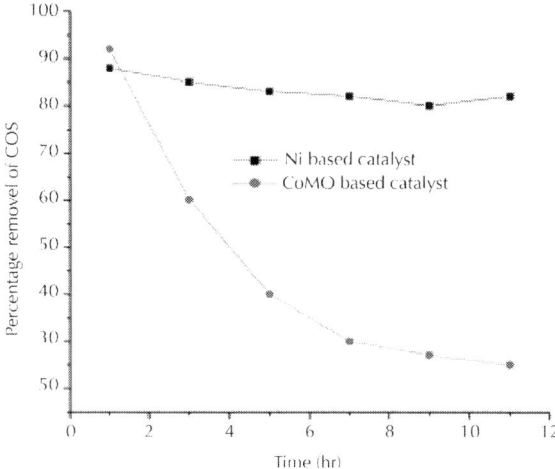

Figure 4: Removal efficiency of COS vs time by the catalysts [28].

The technology for removal of SO_2 for elevated temperature utilizes a dry/semidry calcium based absorbent (lime or limestone) injection into the ductwork after the furnace, or directly into the furnace. Injection of dry/semidry sorbents offers advantages of simplicity and ease of retrofit to existing plants [102].Fig. 5 shows the removal of sulphur using calcium based sorbents.

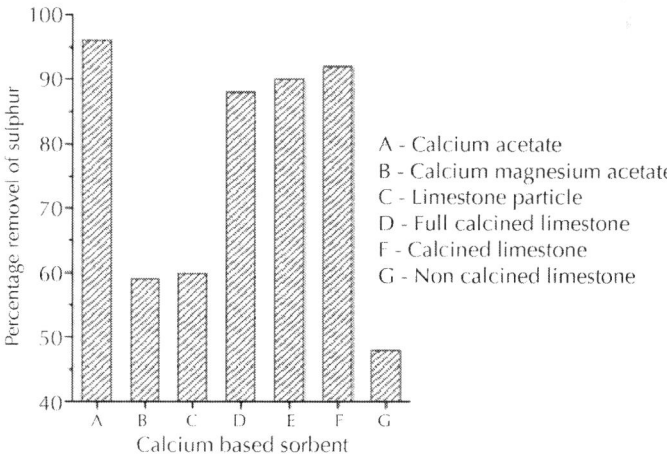

Figure 5: Removal of sulphur using calcium based sorbents [103].

Cold Gas Sulphur Compounds Removal

Low temperature sulphur removal processes employ chemical, physical or mixed solvents. It also utilizes chemical redox process as well as biological processes.

In chemical solvent methods, a liquid solvent is used to create weak chemical bonds between amines and acid gases, commonly H_2S and CO_2. Nowadays many commercial facilities are using three categories of amines (primary, secondary and tertiary) for absorption process. The first commercially available triethanolamine (TEA) is now being replaced by other alkanolamines such as monoethanolamine (MEA), diethanolamine (DEA) and methyldiethanolamine (MDEA). Sulphur in the form of COS is not efficiently removed by chemical solvent methods as it degrades the solvents such as MEA and DEA. Due to that, prior to using these solvents, COS hydrogenation to H_2S is required. Another disadvantage is that solvent has to be replaced continuously due to loss of amines which occurs during operation [104].

Iron based catalysts are being used for LOCAT® process which utilizes slurries of chelated iron and a biocide. The rearation process are given below [105].

$$H_2S + 2Fe^{+++} \rightarrow 2H^+ + S + 2Fe^{++}$$

$$\frac{1}{2}O_2 + H_2O + 2Fe^{++} \rightarrow 2(OH)^- + 2Fe^{+++}$$

The chelate are applied by using venturis. Another method is to bubble the gas stream into auto-circulating tanks of chelate solution [105]. Biological and chemobiological (combination of chemical and biological) processes may also be employed for sulphur removal. Biological processes are generally very slow. Different micro-organisms ranging from photosynthetic autotrophs such as Chlorobiaciae, to chemolithotrophs and autotrophs, such as Thiobacillus denitrificans were studied [106]. As an example for chemobiological approach, the two stage BIO-SR process is commonly used. In the first stage, chemical reaction takes place between ferric sulphate and hydrogen sulphide. In the second step, regeneration of ferric sulphate takes place by biological oxidation using Thiobacillus ferrooxidans.

The technologies for removal of SO_2 at low temperatures use calcium based absorbents using lime/limestone slurry. The wet systems achieve greater than 95% SO_x removal, but require ancillary equipment that amounts to 20% of overall plant costs [102].

Nitrogen Compounds Removal

Nitrogen contaminants in raw fuel gas occur in the form of ammonia (NH_3) and hydrogen cyanide (HCN). They are released from protein structure and heterocyclic aromatic compounds in the feedstock [107]. The concentration of NH_3 forms the dominant part by an order of magnitude. It can either be formed from biomass in the primary stage or from HCN in the secondary gas reactions [108] and [109]. Two-third of NH_3 decomposes to molecular nitrogen at gasification temperature, the rest are H_2 and NO_x. Although CO_2 and N_2 are stable compounds, some unstable oxides like NO and N_2O are also formed in the gasifier [110].

The concentration of NH_3 in fuel gas is usually between several hundred to a few thousand parts per million. The application of fuel gas in gas turbines usually demands ammonia concentrations less than 0.05 mL/L to control nitrogen oxide emissions, but even low value (0.05 µL/L) may still poison catalysts that are used to upgrade fuel gas [111]. Unless and until nitrogen oxides are substantially reduced, the sulphur removal units can also experience problems [92].

Hot Gas Nitrogen Compounds Removal

The hot gas cleaning for nitrogen compounds is primarily focussed on decomposing ammonia rather than removing it from gas streams. However, ammonia released during gasification does not decompose rapidly to achieve low parts per million (ppm) concentration required for many fuel gas applications and thus, selective catalytic oxidation or thermal catalytic oxidation is employed. Catalysts should be chosen in such a way that it selectively oxidizes the nitrogen compounds, thereby avoiding undesired reactions with other gas species.

$$4_{NH3}+6NO \rightarrow 5N_2+6H_2O$$

$$5H_2+NO \rightarrow 3_{NH3}+2H_2O$$

The opposite mechanism of NH_3 formation is used for thermal catalytic decomposition of NH_3. Further, NH_x molecules are dehydrogenated and the N. and H. radicals are then recombined to form N_2 and H_2 [71]. Table 5 gives the summary of typical catalysts for NH_3 decomposition. Nickel and Zeolite are the common catalysts used in thermal catalytic decomposition approach. High NH_3 conversions generally occurs above 500 °C, but higher temperature in the range of 700–800 °C are required to avoid catalyst deactivation from CO induced coking. The Ni based reforming catalysts have proved for 75% reduction of NH_3 when using actual coal derived fuel gas. The only problem is sulphur poisoning which increases with increase in pressure [112]. Tungsten based catalysts such as tungsten carbide (WC) and tungstated zirconia (WZ) are somehow resistant to sulphur and possess good reactivity, physical hardness and multiple types of available active sites [113]. When Ni based catalyst is combined with MnO_3 and Al_2O_3, it overcomes sulphur effects and are capable to remove tar and ammonia simultaneously. This combination proves to be better than 15 other catalysts based on Ni, Fe, Zn–Ti and Cu–Mn. 92% removal efficiency and more than 80% conversion are attained at high H_2S (6 mL/L) concentrations [114]. For NH_3 removal, capability of titanomagnetite was also tested in a hot reactor operated at 500–800 °C under atmospheric pressure using three different gas streams of varying compositions of NH_3 and H_2S in a biomass producer gas. It was found that ferrite (-Fe) was readily formed by the H_2 reduction of titanomagnetite, along with complete decomposition (100%) of NH_3 to N_2 and H_2 [115].

Table 5: Summary of some typical catalysts for NH_3 decomposition [78]

Catalysts	Chemical composition of the catalyst	Gas hourly space velocity (GHSV) (h-1)	Temp (°C)	Feed gas composition	Ammonia conversion (%)
Ru/Al2O3	10 wt% Ru	30,000	450	Pure NH3	32

Ru/SiO2	10 wt% Ru	30,000	400		35
Ru/TiO2	4.8 wt% Ru	30,000	400	Pure NH3	27
Ru/MgO	2.8 wt% Ru	30,000	400	Pure NH3	41
Ni on CeO.9LaO.1O2	10 wt% Ni	100,000	750	500 ppm NH3 in H2, CO, CO2, CH4	97
Ru–Ni/Al2O3	(2–5) wt% Ru–(2–5) wt% Ni	20,000	900	1040 ppm NH3 in 10.5% H2, 28.4% CO, 3.6% CO2, 3.1% H2O, N2 balance	90
Ru/SiO2	10 wt% Ru	30,000	500	Pure NH3	64
Ni/SiO2	10 wt% Ni	30,000	500	Pure NH3	10
Ni/fumed SiO2	5 wt% Ni	30,000	700	Pure NH3	93
Ru/fumed SiO2	5 wt% Ru	30,000	550	Pure NH3	97
Ni/Al2O3	1.2 wt% Ni	30,000	500	15% NH3– 85% He	38
Ni–Ce/Al2O3	1.2 wt%–1% Ce				72
146 (Johnson Matthey)	0.5% Ru/ Al2O3	12,000	500– 700	Pure NH3	7–84
Ni monolith (Ni/ Al2O3)	NA	2500	900	4400 ppm NH3, in 11% CO, 14% CO2, 5% CH4, 10% H2, 12% H2O, 0–500 ppm H2S, 3200 ppm toluene, N2 balance	100
Iron containing dolomite	NA				53
Australian limonite	90 wt% -FeOOH	45,000	500	2000 ppm NH3 in He	99
Australian limonite	90 wt% -FeOOH	45,000	750	2000 ppm NH3 in 20% CO, 10% H2, 3% H2O, He balance	90

Australian limonite	90 wt% -FeOOH	45,000	750	2000 ppm NH3 in 100 ppm H2S, He balance	99
Coal char supported Fe	2–6% wt% Fe	45,000	650	2000 ppm NH3 in He	100

Cold Gas Nitrogen Compounds Removal

The focus of cold gas nitrogen compounds removal is primarily on NH_3 and HCN in fuel gas. They are mostly accomplished by absorption in water, since ammonia is highly water-soluble. Even the condensate of water vapour droplets in fuel gas is capable of removing nitrogen based compounds [116]. Experimentally it is shown that when condensate is formed using a chilled condenser to remove water in fuel gas stream derived from sewage sludge, it results in more than 90% reduction in overall NH_3 content [117]. CO_2 and SO_2 can greatly influence the absorption of ammonia into the aqueous scrubbing medium. A good amount of CO_2 in fact encourages absorption of both ammonia and SO_2 in aqueous phase [118]. Techniques like adsorption and biological processing are being used for cleaning air but in case of fuel gas streams, they show several disadvantages. Activated carbon and zeolites are being used for air purification but it is uneconomical in fuel gas applications. In the end, water scrubbing makes absorption with water the most viable approach for cold gas cleaning of nitrogen removal [119].

Chlorine Compounds Removal

Chlorines in fuel gas are usually present in the form of alkali metal salts which readily vapourizes at elevated temperatures and combines with water vapour to form HCl [120]. Raw fuel gas typically contains chloride in the range of 0.01–0.1 wt% [121]. The concentration of chlorine is critical for a gas engine, because HCl is very corrosive and attacks almost all metal parts of the gas engine and destroys the additives of the lubricating oil. HCl removal is categorized depending on the process temperatures – in dry condition, the fuel gas is made to contact with an adsorbent and at low temperatures, the fuel gas comes in contact with a chemical solution or water.

Hot Gas Chloride Compounds Removal

Hot chlorine gas removal typically employs a sorbent that adsorbs HCl and sometimes alkali [122]. This process adsorbs gaseous HCl to a solid surface at elevated temperatures to generate a salt product through chemisorptions. Between 500 and 550 °C, HCl removal is most effective because of the formation of chemical equilibrium between the gases and solids involved [123]. As temperatures exceed 500 °C, the calcium-based sorbents starts decomposing, thereby decreasing their binding capacity and even releasing adsorbed HCl [123].

Activated carbon, alumina and common alkali oxides in fixed beds are most commonly employed for hot gas cleaning. Alternatives such as alkali-based multioxides can also provide efficiency improvements or environmental benefits, but these are higher in cost than traditional sorbents. Less expensive alternative materials, sodium-rich minerals such as nahcolite, trona, and their derivatives like sodium bicarbonate ($NaHCO_3$) and sodium carbonate (Na_2CO_3) are used for removing HCl at elevated temperatures. Some of the other naturally occurring alternatives include $Ca(OH)_2$ and $Mg(OH)_2$ and their calcined forms like CaO and MgO. Limestone addition in the past has also motivated research for high temperature gasification for using it as inexpensive sorbent for HCl removal [122].

Direct injection of sorbent was also attempted directly into hot gas streams in the temperature range of 600–1000 °C. Calcium-based powders have experimentally showed up to 80% HCl removal in the high temperature range [124]. Fig. 6 shows the effect of catalyst or sorbent type on gaseous HCl formation during gasification of coal at 850 °C.

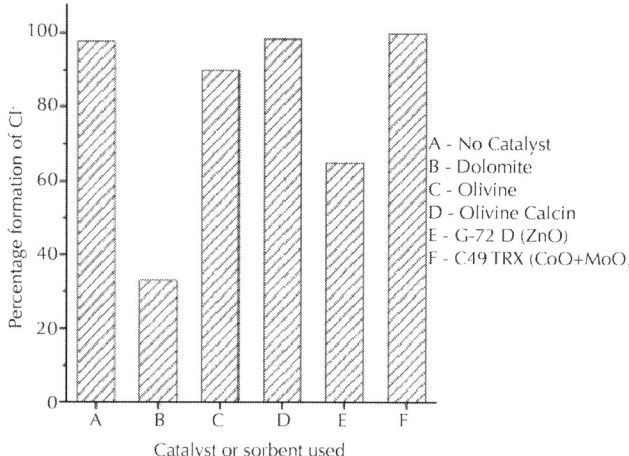

Figure 6: Effect of catalyst or sorbent type on Cl⁻ formation during gasification of coal at 850 °C [125].

Warm Gas Chloride Compounds Removal

For HCl removal by semi-wet removal process, temperature has to be maintained above the point when water condenses. This process utilizes a lime-slurry which is injected with the help of atomization disc. This atomizer atomizes the spray using a rotational disc rather than variation in flow rate. Overall efficiency may be improved by enhancing dispersion and absorption characteristics of the spray even by minimal addition of slurry. As soon as it comes in contact with the gas stream, $Ca(OH)_2$ quickly reacts with HCl to finally form $CaCl_2$ with H_2O. Bag filters are used for the removal of a majority of the particles which are carried along in the gas stream where they are removed in the temperature range of 130–140 °C. Results indicated that Greater than 99.5% HCl and 94% removal of SO_2 was reported in the above process [126]. Adding Mg–Al oxide at 130 °C has been shown to remove 97% of HCl [127].

Cold Gas Chloride Compounds Removal

Chloride removal follows a two process, i.e., deposition of ammonium chloride and absorption of hydrogen chloride vapours. HCl and NH_3

generated in gasification process exist as gases until the fuel gas cools down to about 300 °C and at this temperature, HCl reacts with ammonia in the gas stream to form solid ammonium chloride [17].

$$NH_3(g) + HCl(g) \rightarrow NH_4Cl(s)$$

After the formation, this salt is entrained in the gas flow and the fine particles agglomerate to form larger particles and accumulate on surfaces, leading to fouling of process equipment [17]. Wet scrubbing is very effective in removing major forms of chlorine from the stream. HCl is highly soluble in water, but its removal may be enhanced by sodium carbonate when added to the water [128].

Alkali Metals Removal

Gasification feedstock contains alkali and alkaline earth metals. Concentrations of alkali are greater in coal as compared to biomass [129]. Woody biomass contains more alkaline earth metals, while herbaceous biomass contains greater amount of alkali metals [130]. Alkali metals, principally potassium and to lesser amount sodium naturally exist in biomass. Alkali is both reactive and volatile. It has the capability to melt or even vapourize above 600 °C and makes reactor full of aerosols and vapours [131] and [132]. During combustion, they combine with silicon, sulphur and chlorine resulting in unwanted deposit and corrosion [132] and [133]. In combustion systems, alkali compounds foul heat transfer surfaces, participate in slag formation in grate-fired units and contribute to the formation of fluidized bed agglomerates. This results in reduced efficiency and energy conversion facility, which further increases overall operating cost. In case of integrated gasification combined cycle (IGCC), alkali vapour deposition onto combustion turbine working surface and subsequent hot corrosion are added concerns while dealing with alkali metals [134].

Alkali based catalysts (used for the removal of fuel gas contaminants) as well as transition metal promoters such as cobalt, rubidium, cesium and lithium are also the source of alkali metals. These metals also vapourize at high temperature zones and condense in relatively cooler sections causing corrosion and ash fouling [133] and [137]. Therefore, the level of alkali should be reduced to a few g/kg to a few μg/kg [138].

Hot Gas Alkali Metals Removal

The process to reduce alkali concentrations from fuel gas at higher temperature may be classified into two types: removal via condensation and adsorption on solid sorbent. For condensation process to be effective, temperature below 600 °C are optimum as they also remove the vapours which may bypass particulate removal. However, solid sorbents may be applied even at higher temperatures [120] and [136].

One method of controlling alkali metal vapour in gaseous state is using the alkali getter material. A sorbent used for alkali removal is generally termed as 'getter material'. It should preferably form irreversible adsorptions so that alkali may be retained with getter despite fluctuations in process conditions [139]. In concept, getters are employed as fixed bed of granular inorganic solid sorbents. In gasifier applications, getters are located after hot gas filter equipment to remove alkali vapour before use in a gas turbine. An ideal getter material would possess characteristics such as high temperature capability, rapid rates of adsorption, high loading capacity, transformation of alkali into a less corrosive form, and irreversible adsorption, One mitigation strategy applicable at high temperature is to pass the gas stream through a fixed bed of sorbent or getter material, which preferentially adsorbs alkali via physical adsorption or chemisorption [140].

There are several factors on which sorbent life depends which include the percentage of other contaminants, the temperature of application and the regeneration capability of sorbent. Sorbents also includes different natural minerals like silica, clays or kaolinite. Some synthetic sorbent includes activated alumina which may be synthesized from bauxite minerals. Kaolinite and bauxite are capable of removing alkali metals even at 1000 °C for both in-situ and downstream applications [141] and [142]. Mineral like emathlite is used for low temperature applications because it forms low melting point eutectics with the alkali metals [143]. Bauxite may achieve removal efficiency of 99% in 0.2 s by rapid physical adsorption and it can be regenerated by using boiling water [144]. At 840 °C, activated Al_2O_3 has 98.2% removal efficiency with sodium loading of 6.2 mg/g which, at this temperature, is better than bauxite, kaolinite and clay [145]. Simultaneous removal of alkali and halides may also be done by injecting aluminosilicates and sodium carbonate into the gas downstream [146].

Cold Gas Alkali Metals Removal

When the temperature is lowered to around 300 °C, alkali vapour condenses and agglomerates into small particles. Side-by-side it may also combine with tars, therefore it is simultaneously removed along with particulate and tar in wet scrubbers [147]. Alkali in biomass are water soluble, due to which water washing or leaching is the common approach for removing many alkali compounds [148]. In a study, it was shown that when banagrass (a type of biomass) is mechanically dewatered, rinsed and dewatered again, it removes most of the alkali content and the overall ash is also reduced by 45%. Other contaminants removal percentage were 90% of potassium, 68% of magnesium & sodium, 72% of phosphorous, 98% of chlorine and 55% of sulphur [149]. It is found that washing with acid could also be one of the alternatives. An experiment, conducted prior to pyrolysis of wood waste and wheat straw, shows that 70% of alkali emissions is reduced as compared to only 30% by water wash. Removal rate also depends upon the type of biomass being used. In herbaceous biomass, alkali can easily be removed by water or acid leaching, while woody biomass contains more organically bound alkali, so it is not easily removed by these processes [150].

Heavy Metals Removal

Coal or biomass contains a number of heavy metals such as Ba, Zn, Ni, Cu, Fe, Pb, Mn, Mg and Cd. The environmental pollution due to presence of heavy metals is highly hazardous for plants, animals and human beings and, the toxicities have a direct impact on human health.

Hiraki et al. [151] showed that some appropriate tree species and shrubs in urban areas can greatly reduce the toxic effects of these materials from environment. It was shown that the performance of such trees in comparison to shrubs for refinement of heavy metals such as lead, copper and zinc were found to be more appropriate.

Further, Sorode et al. [152] also compared the leach ability of heavy metals from fly ash, bottom ash, dumping site ash, cement, and brick samples admixtured with fly ash in the area of a thermal power plant.

Research shows that Mg, Mn, and Fe are leached to a larger; Zn, Cu, and Pb to moderate, and Ni to a smaller extent, from the ash samples. The concentrations of Zn, Fe, Mn, Mg and Cd in samples through the leaching from fly ash were generally below the permissible limits of WHO and Indian standards. The leaching of Ni and Pb are slightly higher than WHO permissible limits but below the Indian standards. The admixturing of thermal power plant fly ash in cement and bricks seems to be an eco-friendly practice as far as leaching of heavy metals is concerned.

According to Verwilghen et al. [153] it was seen that the heavy metals of class I such as Ba, Ce, Mg, Mn, Cr are concentrated in the coarse ash (bottom ash), never vapourize, and rapidly undergo a transition to a various solid species and hence they can easily be separated. The class II heavy metals such as Cu, Pb, Se, Zn and As are concentrated in particulate matter while the class III such as Br, Hg, I and Cd are concentrated in the gas phase. An experimental study of the vapourization of class II metals (Cd, Pb and Zn) during thermal treatment of wastes with added heavy metals or phosphate was carried out at laboratory in a pilot scale to evaluate the effects of operating conditions on vapourization rate of metals within wastes and to determine the efficiency of sorbent injection (sodium bicarbonate with activated carbon or hydroxyapatite $Ca_{10}(PO_4)_6(OH)_2$) in flue gas to remove these toxic compounds.

The United States patent (No.: US 8,070,863 B2) provides a flue gas conditioning system for processing an input gas from a low temperature gasification system to an output gas of desired characteristics. As claimed, the system comprises a two-stage process, the first stage is for separating heavy metals and particulate matter in dry phase, and the second stage includes further processing steps of removing acid gases and/or other contaminants. The presence and sequence of processing steps are determined by the composition of the input gas [154].

Preventing the release of mercury from gasifier plants continues to be a challenge. Mercury is often captured through powdered activated carbon injection (ACI) and the capture of the injected carbon particles on a downstream capture device (like electrostatic precipitator or a bag house). An ACI system is relatively simple and inexpensive, consisting of storage equipment, pneumatic conveying system, and injection hardware [155].

Precious metals like palladium (Pd), platinum (Pt), gold (Au), iridium (Ir), and rhodium (Rh), have been used as modifiers for graphite-tube atomic absorption or emission analysis of solid and liquid samples. Among them, Pd has been identified as the best modifier for the adsorption of Hg [156]. Jain et al. [157] theoretically screened potentially high-temperature metal sorbents for Hg capture in fuel gas streams.

Ozaki et al. [158] and Wu et al. [159] and [160] studied the use of iron oxides to remove Hg from synthetic fuel gases. Two oxides, Fe_2O_3 and 1 wt% Fe_2O_3 on TiO_2, were tested in the temperature range of 60–100 °C. After initially higher rates, 5 ppb Hg was removed from the fuel gas at a constant rate of about 60% for both sorbents. It was also found that the presence of H_2S promoted the removal of Hg, while H_2O slowed down Hg removal caused by H_2S.

Dioxins and Furans Removal

Dioxins/furans are highly poisonous and cause severe bronchitis, asthma and choking of lungs in human beings. The formation of poly-chlorinated dibenzodioxins and furans (PCDD/F) compounds in thermal processes is undoubtedly the result of a complex set of competing chemical reactions. Certain operating conditions increase the potential for PCDD/F formation including incomplete combustion of a fuel, an oxidizing atmosphere, presence of a chlorine source, fly ash surfaces acting as a carbon source, fly ash with degenerated graphitic structures, presence of catalytic metals (especially copper, but iron, manganese and zinc could also prove to be a potential catalysts for PCDD/F formation) and temperature range between 250 and 450 °C [159] and [163].

However, these conditions are not satisfied in gasification process and hence, the PCDD/F compounds are least expected to be present in the fuel gas. The reducing environment in gasification process actually precludes the formation of free chlorine from HCl, thereby limiting chlorination of any species in the fuel gas [164] (Table 6).

Table 6: Summary of cleaning technologies for impurities in fuel gas

Sl. no.	Impurities	Cleaning technology	Operating conditions	Reference
1	Solid particles	Cyclones, reverse flow cyclones	Up to 1000 °C	[33]
		Barrier/candle filters, metallic filters, granular bed filters (with Ni and MgO)	Above 400 °C	[36]
			Sintered metallic barrier filter may operate even at 1000 °C	[39]
		Electrostatic precipitator	Between 300–450 °C	[51]
		Spray scrubber, impactor scrubber, venturi scrubber, electrostatic scrubber, etc.	Below 100 °C	[55]
2	Tar	Thermal cracking	1100–1300 °C	[65]
		Catalytic cracking using Fe or Ni based catalysts or olivine or calcined dolomite, etc.	(Ni catalysts are 10 times more effective than dolomite)	[72] and [76]
		Thermal plasma, non-thermal plasmas like pulsed corona, dielectric barrier discharges, DC corona discharges, RF plasma, microwave plasma, etc.	Pulsed corona is effective at 400 °C	[79]
		Nickel based candle filter	Above 850 °C	[86]
		Oil based gas washer (OLGA)	60–450 °C	[76] and [87]
		Venturi scrubber	Below 100 °C	[91]

3	Sulphur	Physical and chemical adsorption, ZnO/FeO sorbents, use of catalysts like CoMo for COS conversion, nickel and iron based catalysts	Mn with V and CuO have desulphurization capacity even above 600 °C	[97]
		Lime/limestone injection, etc.	ZnO sorbent (600–650 °C)	[96]
		LOCAT® process	Below 100 °C	[105]
		Chemobiological process		
4	Nitrogen	Thermal catalytic decomposition of NH3 using Ni and Zeolite as catalysts combined with MnO3 and Al2O3, Tungsten based catalysts such as WC and WZ	700–800 °C	[112]
		Water scrubbing technique	Below 100 °C	[119]
5	Chlorine	Activated carbon, alumina and common alkali oxides	Above 500 °C, the calcium-based sorbents starts decomposing	[123]
		Injection of sorbent like calcium-based powders		
		Alkali-based multioxides such as nahcolite, trona, and their derivatives as sorbent		
		Bag filters	130–140 °C	[126]
		Wet scrubbing to remove solid ammonium chloride formed a low temperature	Below 100 °C	[128]

6	Alkali metals	Condensation and adsorption on solid sorbent, getter material	Kaolinite and bauxite (even at 1000 °C) both for in-situ and downstream applications	[141] and [142]
		Kaolinite and bauxite as adsorbents, minerals like emathlite and bauxite		
		Agglomeration into small particles through condensation and subsequent removal through water washing or leaching	Below 100 °C	[147] and [148]
7	Heavy metals	Shrub refinement, leaching, iron oxides	Pb, Cu and zn were found suitable for shrub refinement	[151]
		Powdered activated carbon injection (ACI), precious metals (Pd, Pt, Au and Rh, sodium bicarbonate with hydroxyapatite	Fe2O3 was found suitable for Hg removal (60–100 °C)	[158] ,[159] and [1 60]

CONCLUSIONS

Fuel gas from gasification may be utilized for power generation after it is properly cleaned in downstream process. The clean fuel gas with high CO and H_2 content has the potential for production of liquid petroleum fuels through Fischer–Tropsch process. However, achieving such clean fuel gas is nowadays a major concern in research. The following observations are made from the studies conducted in this review in order to get quality product gas.

Removal of particulate matters has been achieved to surpass 99.6% using reverse flow cyclones. The fabric filters can also effectively remove particulate matters larger than 1 µm and concentrations less than 1 µg/m³ when operating at temperatures below 250 °C. Olivine may serve as in-situ catalyst when used as bed material in fluidized bed gasifier for tar removal. Nickel catalysts are found to be approximately 10 times more efficient than that of dolomite in decomposing tar at temperature 400–550 °C. Zinc and copper oxides which are found in abundance in nature have the potential for 99% removal of sulphur compounds at above 300 °C. Wet scrubbing techniques is using limestone is mostly preferred at lower temperatures for sulphur removal. Nickel based catalyst, combined with MnO_3 and Al_2O_3, has the capability to remove nitrogen compounds like ammonia and tar simultaneously in hot gas condition. Cold gas cleaning is primarily accomplished by absorption in water, since ammonia is highly soluble in water.

The chlorine compounds in fuel gas, mostly HCl, are removed by using the solid slaked lime and limestone at temperatures in the range of 500–600 °C. At around 300 °C temperature, chloride removal occurs in a two step process, i.e., through deposition of ammonium chloride and absorption of hydrogen chloride vapours. Getter materials are used as sorbents for the removal of alkali metals when the gas temperature is around 600 °C. When the temperature is below 300 °C, the alkali metals may combine with tars and can be simultaneously removed in wet scrubbers. The powdered activated carbon is used nowadays for the removal of mercury. Different techniques such as shrubs refinement, leach ability, hydroxyapatite are useful for complete removal of large number of heavy metals present in fuel gas. Gasification technology employs controlled amount of oxygen for gasification. Side-by-side the temperature inside the gasifier is also low which retards the formation of dioxins and furans.

The methods or techniques as discussed in this review are presently followed for cleaning the dirty fuel gas, but the reduction is only limited to about 70%. Therefore, there are still some scopes to carry out research for complete reduction of gas impurities meeting the challenges in commercialization of such developed technologies. More attention should be given to hot gas cleaning condition as it is more advantageous in terms of thermal efficiency compared to cold gas cleaning. Nevertheless, the steps for gas cleaning process should be reduced further to minimize the thermal loss from the gas stream.

ACKNOWLEDGEMENTS

The authors thankfully acknowledge Director, CSIR - Central Mechanical Engineering Research Institute (CSIR-CMERI), Durgapur, India for his continuous support and encouragement in carrying out this research work.

REFERENCES

1. Ministry of Statistics and Programme Implementation, Energy Statistics, Central Statistics Office, Government of India, 2014. http://www.mospi.gov. in.

2. V.C. Pandey, J.S. Singh, R.P. Singh, N. Singh, M. Yunus, Arsenic hazards in coal fly ash and its fate in Indian scenario, Resour. Conserv. Recycl. 55 (9–10) (2011)819–835, doi:http://dx.doi. org/10.1016/j.resconrec.2011.04.005.

3. B. Buragohain, P. Mahanta, V.S. Moholkar, Biomass gasification for decentralized power generation: the Indian perspective, Renew. Sustainable Energy Rev. 14 (1) (2010)73–92, doi:http:// dx.doi.org/10.1016/j. rser.2009.07.034.

4. J.A. Ruiz, M.C. Juárez, M.P. Morales, P. Muñoz, M.A. Mendívil, Biomass · gasification for electricity generation: review of current technology barriers, 8 Renew. Sustainable Energy Rev. 18 (2013)174–183, doi:http://dx.doi.org/10.1016/j. rser.2012.10.021.

5. http://en.wikipedia.org/wiki/Gasification (date Q4 accessed 08.08.14).

6. G.J. Stiegel, S.J. Clayton, Gasification technologies, A program to Deliver Clean, Secure and Affordable Energy, US Department of Energy, Office of Fossil Energy, National Energy Technology Laboratory, 2001.

7. C. Xu, J. Donald, E. Byambajav, Y. Ohtsuka, Recent advances in catalysts for hot-gas removal of tar and NH3 from biomass gasification, Fuel 89 (8) (2010) 1784–1795, doi:http://dx.doi. org/10.1016/j.fuel.2010.02.014.

8. J. Šulc, J. Štojdl, M. Richter, J. Popelka, K. Svoboda, J. Smetana, J. Vacek, S. Skoblja, P. Buryan, Biomass waste gasification – can

be the two stage process suitable for tar reduction and power generation? Waste Manag. 32 (4) (2012) 692–700, doi:http://dx.doi.org/10.1016/j.wasman.2011.08.015. 21925858.

9. Y. Shen, K. Yoshikawa, Recent progresses in catalytic tar elimination during biomass gasification or pyrolysis—a review, Renew. Sustainable Energy Rev. 21 (2013)371–392, doi:http://dx.doi.org/10.1016/j.rser.2012.12.062.

10. A. Paethanom, S. Nakahara, M. Kobayashi, P. Prawisudha, K. Yoshikawa, Performance of tar removal by absorption and adsorption for biomass gasification, Fuel Process. Technol. 104 (2012)144–154, doi:http://dx.doi.org/ 10.1016/j. fuproc.2012.05.006.

11. K. Chiang, C. Lu, M. Lin, K. Chien, Reducing tar yield in gasification of paper- reject sludge by using a hot-gas cleaning system, Energy 50 (2013)47–53, doi: http://dx.doi.org/10.1016/j. energy.2012.12.010.

12. S.K. Park, J. Ahn, T.S. Kim, Performance evaluation of integrated gasification solid oxide fuel cell/gas turbine systems including carbon dioxide capture, Appl. Energy 88 (9) (2011)2976–2987, doi:http://dx.doi.org/10.1016/j.ape- nergy.2011.03.031.

13. C. Bang-Møller, M. Rokni, B. Elmegaard, Exergy analysis and optimization of a biomass gasification, solid oxide fuel cell and micro gas turbine hybrid system, Energy 36 (8) (2011)4740–4752, doi:http://dx.doi.org/10.1016/j.en- ergy.2011.05.005.

14. N.H. Leibbrandt, A.O. Aboyade, J.H. Knoetze, J.F. Görgens, Process efficiency of bio fuel production via gasification and Fischer–Tropsch synthesis, Fuel 109 (2013)484–492, doi:http://dx.doi.org/10.1016/j.fuel.2013.03.013.

15. K. Kim, Y. Kim, C. Yang, J. Moon, B. Kim, J. Lee, U. Lee, S. Lee, J. Kim, W. Eom, S. Lee, M. Kang, Y. Lee, Long-term operation of biomass-to-liquid systems coupled to gasification and Fischer–Tropsch processes for biofuel production, Bioresour. Technol. 127 (2013)391–399, doi:http://dx.doi.org/10.1016/j.bio-rtech.2012.09.126. 23138062.

16. T.R. Miles, T.R. Miles Jr., L.L. Baxter, R.W. Bryers, B.M. Jenkins, L.L. Oden, Alkali deposits found in biomass power plants, a preliminary investigation of their extent and nature, Summary

Report for the National Renewable Energy Laboratory, NREL, 1995 Subcontract TZ-2-11226-1.

17. P.J. Woolcock, R.C. Brown, A review of cleaning technologies for biomass- derived syngas, Biomass Bioenergy 52 (2013)54–84, doi:http://dx.doi.org/ 10.1016/j.biombioe.2013.02.036

18. H. Hofbauer, ERA-NET-Call gasification – cleaning and treatment of producer gas from biomass gasification, ERA-Net Bioenergy, Institute of Chemical Engineering, Vienna University of Technology, 2015.

19.   S.B. Katole, P. Kumar, R.D. Patil, Environmental Pollutants and Livestock Health: A Review, Veterinary Research International, 2015.

20. R.R. Khan, M.J.A. Siddiqui, Review on effects of particulates; sulfur dioxide and nitrogen dioxide on human health, Int. Res. J. Environ. Sci. 3 (4) (2014) 891 70–73.

21.   S. Martin, W. Griswold, Human health effects of heavy metals, Environmental Science and Technology Briefs for Citizens, Center for Hazardous Substance Research, 2009.

22. W.C. Edwards, D.G. Gregory, Livestock poisoning from oil field drilling fluids, muds and additives, Vet. Hum. Toxicol. 33 (5) (1991) 502–504. 1746149.

23. ATSDR, Toxicological profile of chromium, Agency for Toxic Substances and Drug Registry, US Department of Health and Human Services, Atlanta, GA, 2000.

24. D. Swarup, R.C. Patra, R. Naresh, P. Kumar, P. Shekhar, Blood lead levels in lactating cows reared around polluted localities; transfer of lead in to milk, Sci. Total Environ. 347 (1–3) (2005)106–110,   doi:http://dx.doi.org/10.1016/j.   scitotenv.2004.12.055. 16084971.

25. A. Hoffmann, L. Stein, P. Bradshaw, Gas cyclones and swirl tubes: principles, design, and operation, Appl. Mech. Rev. 56 (2) (2003) B28–B29, doi:http://dx. doi.org/10.1115/1.1553446.

26. K. Szemmelveisz, I. Szács, ÁB. Palotás, L. Winkler, E.G. Eddings, Examination of the combustion conditions of herbaceous biomass, Fuel Process. Technol. 90 (6) (2009), doi:http://dx.doi. org/10.1016/j.fuproc.2009.03.001. 839e47.

27. E. Gustafsson, M. Strand, M. Sanati, Physical and chemical characterization of aerosol particles formed during the thermochemical conversion of wood pellets using a bubbling fluidized bed gasifier, Energy Fuels 21 (6) (2007), doi: http://dx.doi.org/10.1021/ef7002552. 3660e7.

28. B. Dou, C. Wang, H. Chen, Y. Song, B. Xie, Y. Xu, C. Tan, Research progress of hot gas filtration, desulphurization and HCl removal in coal-derived fuel gas: a review, Chem. Eng. Res. Des. 90 (11) (2012)1901–1917, doi:http://dx.doi.org/ 10.1016/j. cherd.2012.04.009.

29. R.A. Newby, W. Yang, R.L. Bannister, Fuel gas cleanup parameters in air-blown IGCC, J. Eng. Gas Turbines Power 122 (2) (2000)247–254, doi:http://dx.doi. org/10.1115/1.483202.

30. G.J. Stiegel, R.C. Maxwell, Gasification technologies: the path to clean, affordable energy in the 21st century, Fuel Process. Technol. 71 (1–3) (2001) 79–97, doi:http://dx.doi.org/10.1016/ S0378-3820 (01)00138-2.

31. J.P.K. Seville, 1st ed., Gas Cleaning in Demanding Applications, vol. xv, Blackie Academic & Professional, London, New York, 1997, pp. 308.

32. C. Higman, M. van der Burgt, 2nd ed., Gasification, vol. xvi, Gulf Professional, Elsevier Science, Amsterdam, Boston, 2008, pp. 435.

33. W. Koch, W. Licht, New design approach boosts cyclone efficiency, Chem. Eng. 84 (24) (1977). 80e8.

34. R.L. Salcedo, M.J. Pinho, Pilot- and industrial-scale experimental investigation of numerically optimized cyclones, Ind. Eng. Chem. Res. 42 (1) (2003), doi:http://dx.doi.org/10.1021/ie020195e. 145e54.

35. K. Lee, J. Sohn, Y. Park, Filtration performance characteristics of ceramic candle filter based on inlet structure of high-temperature and high-pressure dust collectors, J. Ind. Eng. Chem. 21 (2015)101–110, doi:http://dx.doi.org/ 10.1016/j.jiec.2014.09.004.

36. S.D. Sharma, M. Dolan, D. Park, L. Morpeth, A. Ilyushechkin, K. McLennan, D. J. Harris, K.V. Thambimuthu, A critical review of syngas cleaning technologies — fundamental limitations and practical problems, Powder Technol. 180 (1–2) (2008)115–121, doi:http://dx.doi.org/10.1016/j.pow-tec.2007.03.023.

37. S.D. Sharma, M. Dolan, A.Y. Ilyushechkin, K.G. McLennan, T. Nguyen, D. Chase, Recent developments in dry hot syngas cleaning processes, Fuel 89 (4) (2010), doi:http://dx.doi.org/10.1016/j.fuel.2009.05.026. 817e26.

38. S. Desu, C.H. Peng, Inventors, Center for Innovative Technology, Assignee, Coating Porous Materials with Metal Oxides and Other Ceramics by Metal Organic Chemical Vapor Deposition (MOCVD), United States of America, Patent 5,262,199 (1993).

39. I. Schildermans, J. Baeyens, Reviewing the Growing Potential of Porous Sintered Metal Filters, Prague, Czech Republic: Czech Society of Chemical Engineering [p. Hexion Specialty Chemicals; Mitsubishi Chemical Corporation; CS Cabot; Zentiva; BorsodChem MCHZ] (2006).

40. B. Gardner, X. Guan, R.A. Martin, J. Spain, Hot gas filtration meeting turbine requirements for particulate matter, RenoTahoe, American Society of Mechanical Engineers, NV, United States, 2005, pp. 439–451.

41. D. Stanghelle, T. Slungaard, O.K. Sønju, Granular bed filtration of high temperature biomass gasification gas, J. Hazard. Mater. 144 (3) (2007) 668e72, doi:http://dx.doi.org/10.1016/j.jhazmat.2007.01.092. 17337119.

42. J.A. Ritzert, R.C. Brown, J. Smeenk, Filtration efficiency of a moving bed granular filter, Science in Thermal and Chemical Biomass Conversion, Victoria, Victoria, BC, Canada, 2004.

43. R.C. Brown, H. Shi, G. Colver, S. Soo, Similitude study of a moving bed granular filter, Powder Technol. 138 (2–3) (2003)201–210, doi:http://dx.doi.org/ 10.1016/j.powtec.2003.09.002.

44. . Smid, S. Hsiau, C. Peng, H. Lee, Moving bed filters for hot gas cleanup, Filtr. Sep. 42 (6) (2005)34–37, doi:http://dx.doi.org/10.1016/S0015-1882(05) 70591-7.

45. J. Smid, S.S. Hsiau, C.Y. Peng, H.T. Lee, Hot gas cleanup: new designs for moving bed filters, Filtr. Sep. 42 (10) (2005) 36–39.

46. I.A. El-Hedok, L. Whitmer, R.C. Brown, The influence of granular flow rate on the performance of a moving bed granular filter, Powder Technol. 214 (1) (2011), doi:http://dx.doi.org/10.1016/j.powtec.2011.07.037. 69e76.

47. L. Ma, G.V. Baron, Mixed zirconia–alumina supports for Ni/MgO based catalytic filters for biomass fuel gas cleaning, Powder Technol. 180 (1–2) (2008)21–29, doi:http://dx.doi.org/10.1016/j.powtec.2007.02.035.

48. S. Rapagnà, K. Gallucci, M. Di Marcello, P.U. Foscolo, M. Nacken, S. Hei denreich, In situ catalytic ceramic candle filtration for tar reforming and particulate abatement in a fluidized-bed biomass gasifier, Energy Fuels 23 (7) (2009)3804–3809, doi:http://dx.doi.org/10.1021/ef900166t.

49. D.A. Lloyd, Electrostatic Precipitator Handbook, vol. xiv, A. Hilger, Bristol, England; Philadelphia, 1988, pp. 239.

50. J.R. McDonald, A.H. Dean, Electrostatic Precipitator Manual, vol. xi, Noyes Data Corp, Park Ridge, NJ, 1982.

51. R.F. Probstein, R.E. Hicks, Synthetic Fuels, vol. xiv, Dover Publications, Mineola, NY, 2006, pp. 490.

52. A.J.M. Pemen, S.A. Nair, K. Yan, E.J.M. van Heesch, K.J. Ptasinski, A.A.H. Drinkenburg, Pulsed corona discharges for tar removal from biomass derived fuel gas, Plasmas Polym. 8 (3) (2003). 209e24.

53. W. Peukert, High temperature filtration in the process industry, Filtr. Sep. 35 (5) (1998)461–464, doi:http://dx.doi.org/10.1016/S0015-1882(98)80015-3.

54. P. Hasler, R. Buehler, Th. Nussbaumer, Biomass for energy and industry, 10th European Conference and Technology Exhibition, Wurzburg, Germany, 1998.

55. K.C. Schifftner, H.E. Hesketh, 2nd ed., Wet Scrubbers, vol. xv, Technomic, Lancaster, PA, 1996, pp. 206.

56. L.C. Laurence, D. Ashenafi, Syngas treatment unit for small scale gasification – application to IC engine gas quality requirement, J. Appl. Fluid Mech. 5 (1) (2012) 95–103.

57. T.A. Milne, R.J. Evans, N. Abatzoglou, Biomass Gasifier "Tars": Their Nature, Formation, and Conversion. Report No.: NREL/TP-570-25357, NREL, Golden, Colorado, USA, 1998.

58. K. Maniatis, A.A.C.M. Beenackers, Introduction: tar protocols, IEA gasification tasks, Biomass Bioenergy 18 (1) (2000)1–4, doi:http://dx.doi.org/10.1016/ S0961-9534(99)00072-0.

59. S. Anis, Z.A. Zainal, Tar reduction in biomass producer gas via mechanical, catalytic and thermal methods: a review, Renew. Sustainable Energy Rev. 15 (5) (2011)2355–2377, doi:http://dx.doi.org/10.1016/j.rser.2011.02.018.

60. M. Houben, H. Delange, A. Vansteenhoven, Tar reduction through partial combustion of fuel gas, Fuel 84 (7–8) (2005)817–824, doi:http://dx.doi.org/ 10.1016/j.fuel.2004.12.013.

61. M. Van't Hoff, Advanced gasification, gas cleaning and product gas utilization, Dahlman, New horizons in gasification, 12th European Gasification Conference, 10–13 March Rotterdam, The Netherlands, 2014.

62. P.C.A. Bergman, S.V.B. van Paasen, H. Boerrigter, The novel "OLGA" technology 970 for complete tar removal from biomass producer gas, Pyrolysis and Gasification of Biomass and Waste, ECN, Strasbourg, France, 2002.

63. E.G. Baker, M.D. Brown, R.H. Moore, L.K. Mudge, D.C. Elliott, Engineering analysis of biomass gasifier product gas cleaning technology, Battelle Memorial Institute, Biofuels and Municipal Waste Technology Division, Pacific, Northwest, Richland, Washington, 1986.

64. C. Van der meijden, Biomass and waste gasification, gas cleaning and options to use the fuel gas, Biomass Energy Engineering, Ecoremed Workshop, Naples, 2013.

65. A. Tregrossi, A. Ciajolo, R. Barbella, The combustion of benzene in rich premixed flames at atmospheric pressure, Combust. Flame 117 (3) (1999), doi:http://dx.doi.org/10.1016/S0010-2180 (98)00157-6. 553e61.

66. J. Fjellerup, J. Ahrenfeldt, U. Henriksen, B. Gobel, Formation, decomposition, and cracking of biomass tars in gasification, Agency DE, Technical University of Denmark, 2005, pp. 60 Report No.: MEK-ET-2005-05.

67. P. Brandt, U. Henriksen, Decomposition of tar in gas from updraft gasifier by thermal cracking, 1st World Conference on Biomass for Energy and Industry, James and James, Sevilla, Spain, 2000, pp. 1756–1758.

68. D. Sutton, B. Kelleher, J.R.H. Ross, Catalytic conditioning of organic volatile products produced by peat pyrolysis, Biomass

Bioenergy 23 (3) (2002), doi: http://dx.doi.org/10.1016/S0961-9534(02)00041-7. 209e16.

69. A. Bosmans, S. Wasan, L. Helsen, Waste to clean syngas: avoiding tar problems, 2nd International Enhanced Landfill Mining Symposium, Houthalen-Helchteren, 2013.

70. P. Simell, Tar decomposition by novel catalytic hot gas cleaning methods, Bioenergy Enlarged Perspectives, NOVACAT, Budapest, 2003.

71. A. Steynberg, M. Dry, Fischer–Tropsch Technology, vol. xxi, Elsevier, Amsterdam, Boston, 2004, pp. 700.

72. Z. Abu El-Rub, E.A. Bramer, G. Brem, Review of catalysts for tar elimination in biomass gasification processes, Ind. Eng. Chem. Res. 43 (22) (2004), doi: http://dx.doi.org/10.1021/ie0498403. 6911e9.

73. M.L. Mastellone, U. Arena, Olivine as a tar removal catalyst during fluidized bed gasification of plastic waste, AIChE J. 54 (6) (2008), doi:http://dx.doi.org/ 10.1002/aic.11497. 1656e67.

74. Z. Zhao, N. Lakshminarayanan, S.L. Swartz, G.B. Arkenberg, L.G. Felix, R.B. Slimane, C.C. Choi, U.S. Ozkan, Characterization of olivine-supported nickel silicate as potential catalysts for tar removal from biomass gasification, Appl. Catal. A 489 (2015)42–50, doi:http://dx.doi.org/10.1016/j. apcata.2014.10.011.

75. Z. Abu El-Rub, E.A. Bramer, G. Brem, Experimental comparison of biomass chars with other catalysts for tar reduction, Fuel 87 (10–11) (2008) 2243–2252, doi:http://dx.doi.org/10.1016/j. fuel.2008.01.004.

76. F.L. Chan, A. Tanksale, Review of recent developments in Ni-based catalysts for biomass gasification, Renew. Sustainable Energy Rev. 38 (2014)428–438, doi:http://dx.doi.org/10.1016/j. rser.2014.06.011.

77. K. Engelen, Y. Zhang, D.J. Draelants, G.V. Baron, A novel catalytic filter for tar removal from biomass gasification gas: improvement of the catalytic activity in presence of H2S, Chem. Eng. Sci. 58 (3–6) (2003)665–670, doi:http://dx. doi.org/10.1016/S0009-2509(02)00593-6.

78. C. Xu, J. Donald, E. Byambajav, Y. Ohtsuka, Recent advances in catalysts for hot-gas removal of tar and NH3 from biomass

gasification, Fuel 89 (2010) 1784–1795, doi:http://dx.doi.org/10.1016/j.fuel.2010.02.014.

79.   S.A. Nair, A.J.M. Pemen, K. Yan, F.M. van Gompel, H.E.M. van Leuken, E.J.M. van Heesch, K.J. Ptasinski, A.A.H. Drinkenburg, Tar removal from biomass-derived fuel gas by pulsed corona discharges, Fuel Process. Technol. 84 (1–3) (2003) 161–173, doi:http://dx.doi.org/10.1016/S0378-3820(03)00053-5.

80.   R.P. Gupta, B.S. Turk, J.W. Portzer, D.C. Cicero, Desulfurization of syngas in a transport reactor, Environ. Progress 20 (3) (2001), doi:http://dx.doi.org/ 10.1002/ep.670200315. 187e95.

81.   P. Hasler, T. Nussbaumer, Gas cleaning for IC engine applications from fixed bed biomass gasification, Biomass Bioenergy 16 (6) (1999)385–395, doi: http://dx.doi.org/10.1016/S0961-9534(99)00018-5.

82.   E. Simeone, M. Siedlecki, M. Nacken, S. Heidenreich, W. de Jong, High temperature gas filtration with ceramic candles and ashes characterisation during steam–oxygen blown gasification of biomass, Fuel 108 (2013)99–111, doi:http://dx.doi.org/10.1016/j.fuel.2011.10.030.

83.   F. Lettner, H. Timmerer, P. Haselbacher, Biomass gasification – state of the art description, Guideline for Safe and Eco-Friendly Biomass Gasification Intelligent Energy – Europe (IEE), Graz University of Technology – Institute of Thermal Engineering, Austria, 2007.

84.   J. Han, H. Kim, The reduction and control technology of tar during biomass gasification/pyrolysis:an overview, Renew. Sustainable Energy Rev. 12 (2) (2008)397–416, doi:http://dx.doi.org/10.1016/j.rser.2006.07.015.

85.   S. Rapagnà, K. Gallucci, M. Di Marcello, M.A. Matt, M. Nacken, S. Heidenreich, P.U. Foscolo, Gas cleaning, gas conditioning and tar abatement by means of a catalytic filter candle in a biomass fluidized-bed gasifier, Bioresour. Technol. 101 (18) (2010)7123–7130, doi:http://dx.doi.org/10.1016/j.bio- rtech.2010.04.019. 20434907.

86.   D.J. Draelants, H. Zhao, G.V. Baron, Catalytic conversion of tars in biomass gasification fuel gases with nickel-activated ceramic filters, in: A. Corma, F.V. Melo, S. Mendioroz, J.L.G. Fierro (Eds.),

Studies in Surface Science and Ca- talysis, vol. 130, Elsevier Science B.V., 2000, pp. 1595–1600.

87. R.W.R. Zwart, A. Bos, J. Kuipers, Principle of OLGA Tar Removal System, Energy Research Centre of the Netherlands (ECN), 2010, pp. 2.

88. B.V. Drift, Gasification: Gas Cleaning and Gas Conditioning, ECN, Newcastle, UK, 2013.

89. S.V.B. Van Paasen, L.P.L.M. Rabou, R. Bar, Tar removal with a wet electrostatic precipitator (ESP): a parametric study, 2nd World Conference and Technology Exhibition on Biomass for Energy, Industry and Climate Protection 1032 (2004) 33–39.

90. .D. Bapat, Application of ESP for gas cleaning in cement industry – with reference to India, J. Hazard. Mater. 81 (3) (2001)285–308, doi:http://dx.doi. org/10.1016/S0304-3894(00)00352-6. 11163692.

91. P. Hasler, T. Nussbaumer, R. Buehler, Evaluation of Gas Cleaning Technologies for Small Scale Biomass Gasifiers, Swiss Federal Office of Energy and Swiss Federal Office for Education and Science, 1997.

92. R. Lovell, S. Dylewski, C. Peterson, Control of Sulfur Emissions from Oil Shale Retorts, EPA, Cincinnati, Ohio, 1981, pp. 190 Report No.: EPA 600/7-82-016.

93. J.B. Young, J.R. Tabberer, J.E. Fackrell, Deposition of corrosive alkali salt vapors on the blades of gas turbines fuelled by coal derived fuel gases, J. Eng. Gas Turbines Power 135 (9) (2013) 13.

94. L.K. Wang, N.C. Pereira, Y.T. Hung, Desulfurization and emissions control, Advanced Air and Noise Pollution Control, vol. 2(2005) . http://www. springer.com/978-1-58829-359-6.

95. A. Steynberg, M. Dry, Fischer–Tropsch Technology, vol. xxi, Elsevier, Amsterdam, Boston, 2004.

96. D.Vamvuka, C.Arvanitidis, D.Zachariadis, Flue gas desulfurization at high temperatures: a review, Environ. Eng. Sci. 21 (4) (2004), doi:http://dx.doi. org/10.1089/1092875041358557. 525e47.

97. W. Torres, S.S. Pansare, J.G. Goodwin, Hot gas removal of tars, ammonia, and hydrogen sulfide from biomass gasification gas, Catal. Rev. 49 (4) (2007), doi: http://dx.doi. org/10.1080/01614940701375134. 407e56.

98. S.S. Tamhankar, M. Bagajewicz, G.R. Gavalas, P.K. Sharma, M. Flytzani-Ste- phanopoulos, Mixed-oxide sorbents for high-temperature removal of hydrogen sulfide, Ind. Eng. Chem. Proc. Des. Dev. 25 (2) (1986), doi:http://dx. doi.org/10.1021/i200033a014. 429e37.

99. J.M. Sánchez-Hervás, J. Otero, E. Ruiz, A study on sulphidation and regeneration of Z-Sorb III sorbent for H2S removal from simulated ELCOGAS IGCC syngas, Chem. Eng. Sci. 60 (11) (2005)2977–2989, doi:http://dx.doi.org/ 10.1016/j.ces.2005.01.018.

100. Y. Ohtsuka, N. Tsubouchi, T. Kikuchi, H. Hashimoto, Recent progress in Japan on hot gas cleanup of hydrogen chloride, hydrogen sulfide and ammonia in coal-derived fuel gas, Powder Technol. 190 (3) (2009)340–347, doi:http://dx. doi.org/10.1016/j. powtec.2008.08.012.

101. M. Kawase, M. Otaka, Removal of H2S using molten carbonate at high temperature, Waste Manag. 33 (12) (2013)2706–2712, doi:http://dx.doi.org/ 10.1016/j.wasman.2013.08.002. 24035726.

102. S.S. Tamhankar, M. Bagajewicz, G.R. Gavalas, P.K. Sharma, M. Flytzani-Stephanopoulos, Mixed-oxide sorbents for high temperature removal of hydrogen-sulfide, Ind. Eng. Chem. Proc. Des. Dev. 25 (2) (1986) 429–437.

103. X. Meng, W. de Jong, R. Pal, A.H.M. Verkooijen, In bed and downstream hot gas desulphurization during solid fuel gasification: a review, Fuel Process. Technol. 91 (8) (2010)964–981, doi:http://dx.doi.org/10.1016/j. fuproc.2010.02.005.

104. N. Korens, D.R. Simbeck, D.J. Wilhelm, Process Screening Analysis of Alternative Gas Treating Sulfur Removal for Gasification, SFA Pacific, Mountain View, California, 2002.

105. J. Watson, K.D. Jones, T. Barnette, Remove hydrogen sulfide from fuel gas, Hydrocarbon Process. 87 (1) (2008) 81–84.

106. A.B. Jensen, C. Webb, Treatment of H2S-containing gases: a review of microbiological alternatives, Enzyme Microb. Technol. 17 (1) (1995), doi: http://dx.doi.org/10.1016/0141-0229(94)00080-B. 2e10.

107. K. Hansson, J. Samuelsson, C. Tullin, L. Åmand, Formation of HNCO, HCN, and NH3 from the pyrolysis of bark and nitrogen-

containing model compounds, Combust. Flame 137 (3) (2004), doi:http://dx.doi.org/10.1016/j.combust-   flame.2004.01.005. 265e77.

108. J.F. Espinal, T.N. Truong, F. Mondragón, Mechanisms of NH3 formation during the reaction of H2 with nitrogen containing carbonaceous materials, Carbon 45 (11) (2007), doi:http://dx.doi.org/10.1016/j.carbon.2007.06.011. 2273e9.

109. M. Becidan, Ø Skreiberg, J.E. Hustad, NOx and N2O precursors (NH3 and HCN) in pyrolysis of biomass residues, Energy Fuels 21 (2) (2007), doi:http://dx.doi. org/10.1021/ef060426k. 1173e80.

110. G. Busca, C. Pistarino, Abatement of ammonia and amines from waste gases: a summary, J. Loss Prev. Proc. Ind. 16 (2) (2003), doi:http://dx.doi.org/10.1016/ 1076   S0950-4230(02)00093-1. 157e63.

111. J. Zhou, S.M. Masutani, D.M. Ishimura, S.Q. Turn, C.M. Kinoshita, Release of fuel-bound nitrogen in biomass during high temperature pyrolysis and gasification, 32nd Intersociety Energy Conversion Engineering Conference (1997) 1785–1790.

112. W. Mojtahedi, M. Ylitalo, T. Maunula, J. Abbasian, Catalytic decomposition of ammonia in fuel gas produced in pilotscale pressurized fluidized-bed gasifier, Fuel Process. Technol. 45 (3) (1995) 221–236.

113. S.S. Pansare, J.G. Goodwin, S. Gangwal, Simultaneous ammonia and toluene decomposition on tungsten-based catalysts for hot gas cleanup, Ind. Eng. Chem. Res. 47 (22) (2008) 8602–8611.

114. W. Mojtahedi, J. Abbasian, Catalytic decomposition of ammonia in a fuel gas at high temperature and pressure, Fuel 74 (11) (1995), doi:http://dx.doi.org/ 10.1016/0016-2361(95)00152-U. 1698e703.

115. J. Hongrapipat, A.C.K. Yip, A.T. Marshall, W.L. Saw, S. Pang, Investigation of simultaneous removal of ammonia and hydrogen sulphide from producer gas in biomass gasification by titanomagnetite, Fuel 135 (2014)235–242, doi: http://dx.doi.org/10.1016/j.fuel.2014.06.037.

116. T. Proll, I.G. Siefert, A. Friedl, H. Hofbauer, Removal of NH3 from biomass gasification producer gas by water condensing in an organic solvent scrubber, Ind. Eng. Chem. Res. 44 (5) (2005) 1576–1584.

117. F. Pinto, H. Lopes, R.N. André, M. Dias, I. Gulyurtlu, I. Cabrita, Effect of experimental conditions on gas quality and solids produced by sewage sludge cogasification. 1: sewage sludge mixed with coal, Energy Fuels 21 (5) (2007)2737–2745, doi:http://dx.doi.org/10.1021/ef0700836.

118. H. Bai, A.C. Yeh, Removal of CO2 greenhouse gas by ammonia scrubbing, Ind. Eng. Chem. Res. 36 (6) (1997), doi:http://dx.doi.org/10.1021/ie960748j. 2490e3.

119. H. Fortier, P. Westreich, S. Selig, C. Zelenietz, J.R. Dahn, Ammonia, cyclo hexane, nitrogen and water adsorption capacities of an activated carbon impregnated with increasing amounts of ZnCl2, and designed to chemisorb gaseous NH3 from an air stream, J. Colloid Interface Sci. 320 (2) (2008) 423–435.

120. C. Higman, M. van der Burgt, 2nd ed., Gasification, vol. xvi, Gulf Professional, Elsevier Science, Amsterdam, Boston, 2008, pp. 435.

121. R.M. Davidson, Chlorine and Other Halogens in Coal, IEAPER/28, IEA Coal Research, London, 1996.

122. B. Dou, B. Chen, J. Gao, X. Sha, HCl removal and chlorine distribution in the mass transfer zone of a fixed-bed reactor at high temperature, Energy Fuels 20 (3) (2006), doi:http://dx.doi.org/10.1021/ef060018g. 959e63.

123. B. Dou, J. Gao, X. Sha, A study on the reaction kinetics of HCl removal from high-temperature coal gas, Fuel Process. Technol. 72 (1) (2001), doi:http://dx. doi.org/10.1016/S0378-3820(01)00176-X. 23e33.

124. B. Shemwell, Y.A. Levendis, G.A. Simons, Laboratory study on the high- temperature capture of HCl gas by dry-injection of calcium-based sorbents, Chemosphere 42 (5–7) (2001)785–796, doi:http://dx.doi.org/10.1016/   S0045-6535(00)00252-6. 11219704.

125. F. Pinto, H. Lopes, R.N. André, I. Gulyurtlu, I. Cabrita, Effect of catalysts in the quality of syngas and by-products obtained by co-gasification of coal and wastes. 2: heavy metals, sulphur and halogen compounds abatement, Fuel 87 (7) (2008)1050–1062, doi:http://dx.doi.org/10.1016/j.fuel.2007.06.014.

126. T. Spooren, A. Raveel, B. Adams, G. Du Toit, P. Waller, Semiwet scrubbing: design and operational experience of a state-of-

the-art unit, Environ. Progress 25 (3) (2006), doi:http://dx.doi. org/10.1002/ep.10138. 201e7.

127. T. Kameda, N. Uchiyama, K.S. Park, G. Grause, T. Yoshioka, Removal of hydrogen chloride from gaseous streams using magnesium–aluminum oxide, Chemosphere 73 (5) (2008)844e7, doi:http://dx.doi.org/10.1016/j.   chemosphere.2008.06.022. 18649922.

128. A.L. Kohl, R. Nielsen, 5th ed., Gas purification, vol. viii, Gulf Publishing, Houston, TX, 1997, pp. 1395.

129. M.P. Glazer, N.A. Khan, W. de Jong, H. Spliethoff, H. Schürmann, P. Monkhouse, Alkali metals in circulating fluidized bed combustion of biomass and coal: measurements and chemical equilibrium analysis, Energy Fuels 19 (5) (2005), doi:http:// dx.doi.org/10.1021/ef0500336. 1889e97.

130. A.A. Khan, W. de Jong, P.J. Jansens, H. Spliethoff, Biomass combustion in fluidized bed boilers: potential problems and remedies, Fuel Process. Technol. 90 (1) (2009), doi:http://dx.doi. org/10.1016/j.fuproc.2008.07.012. 21e50.

131. J. Smeenk, R.C. Brown, D. Eckels, Determination of vapor phase alkali content during biomass gasification, Fourth Biomass Conference of the Americas, Oakland, CA, 1999.

132. D.C. Dayton, R.J. French, T.A. Milne, Direct observation of alkali vapor release during biomass combustion and gasification. 1: application of molecular beam/mass spectrometry to switchgrass combustion, Energy Fuels 9 (5) (1995), doi:http://dx.doi. org/10.1021/ef00053a018. 855e65.

133. S. Turn, C.M. Kinoshita, D.M. Ishimura, J. Zhou, The fate of inorganic constituents of biomass in fluidized bed gasification, Fuel 77 (3) (1998), doi: http://dx.doi.org/10.1016/S0016-2361(97)00190-7. 135e46.

134. Z. Mei, Z. Shen, Q. Zhao, W. Wang, Y. Zhang, Removal and recovery of gas- phase element mercury by metal oxide-loaded activated carbon, J. Hazard. Mater. 152 (2) (2008)721–729, doi:http://dx.doi.org/10.1016/j.jhaz- mat.2007.07.038. 17765397.

135. J. Smeenk, R.C. Brown, Evaluation of an integrated gasification/ fuel cell power plant, Third Biomass Conference of the Americas Montreal, Canada, 1997.

136. J.P. Trembly, R.S. Gemmen, D.J. Bayless, The effect of IGFC warm gas cleanup system conditions on the gas–solid partitioning and form of trace species in coal fuel gas and their interactions with SOFC anodes, J. Power Sources 163 (2) (2007) 986–996.

137. T. Valmari, T.M. Lind, E.I. Kauppinen, G. Sfiris, K. Nilsson, W. Maenhaut, Field study on ash behavior during circulating fluidized-bed combustion of biomass. 2: ash deposition and alkali vapor condensation, Energy Fuels 13 (2) (1998) 390–395.

138. A. Kling, C. Andersson, A. Myringer, D. Eskilsson, S. Jaras, Alkali deactivation of high-dust SCR catalysts used for NOx reduction exposed to flue gas from 100 MW-scale biofuel and peat fired boilers: influence of flue gas composition, Appl. Catal. B. Environ. 69 (3–4) (2007)240–251, doi:http://dx.doi.org/ 10.1016/j. apcatb.2006.03.022.

139. W.A. Punjak, M. Uberoi, F. Shadman, Control of ash deposition through the high temperature adsorption of alkali vapors on solid sorbents, Symposium on Ash Deposition, 197th Annual meeting of the American Chemical Society, University of Arizona, Dallas, TX, 1989.

140. S. Turn, C. Kinoshita, D. Ishimura, J. Zhou, Gas Purification, National Renewable Energy Laboratory, 1617 Cole Boulevard Golden, Colorado 80401- 3393, 2000 NREL/SR-570-26160.

141. J.P.K. Seville, first ed., Gas Cleaning in Demanding Applications, vol. xv, Blackie Academic & Professional, London, New York, 1997.

142. B. Dou, W. Pan, J. Ren, B. Chen, J. Hwang, T. Yu, Single and combined removal of HCl and alkali metal vapor from high-temperature gas by solid sorbents, Energy Fuels 21 (2) (2007), doi:http://dx.doi.org/10.1021/ef060266c. 1019e23.

143. W.A. Punjak, M. Uberoi, F. Shadman, High-temperature adsorption of alkali vapors on solid sorbents, AIChE J. 35 (7) (1989), doi:http://dx.doi.org/ 10.1002/aic.690350714. 1186e94.

144. K. Cummer, R.C. Brown, Ancillary equipment for biomass gasification, Biomass Bioenergy 23 (2) (2002), doi:http://dx.doi. org/10.1016/S0961-9534 (02)00038-7. 113e28.

145. B. Dou, W. Shen, J. Gao, X. Sha, Adsorption of alkali metal vapor from high- temperature coal-derived gas by solid sorbents, Fuel Process. Technol. 82 (1) (2003), doi:http://dx.doi.org/10.1016/ S0378-3820(03)00027-4. 51e60.

146. S.D. Sharma, M. Dolan, A.Y. Ilyushechkin, K.G. McLennan, T. Nguyen, D. Chase, Recent developments in dry hot syngas cleaning processes, Fuel 89 (4) (2010), doi:http://dx.doi.org/10.1016/j.fuel.2009.05.026. 817e26.

147. O. Hirohata, T. Wakabayashi, K. Tasaka, C. Fushimi, T. Furusawa, P. Kuchonthara, A. Tsutsumi, Release behavior of tar and alkali and alkaline earth metals during biomass steam gasification, Energy Fuels 22 (6) (2008), doi: http://dx.doi.org/10.1021/ef800390n. 4235e9.

148. L.L. Baxter, T.R. Miles, T.R. Miles, B.M. Jenkins, T. Milne, D. Dayton, R.W. Bryers, L.L. Oden, The behavior of inorganic material in biomass-fired power boilers: field and laboratory experiences, Fuel Process. Technol. 54 (1–3) (1998) 47–78, doi:http://dx.doi.org/10.1016/S0378-3820(97)00060-X.

149. S.Q. Turn, C.M. Kinoshita, D.M. Ishimura, Removal of inorganic constituents of biomass feedstocks by mechanical dewatering and leaching, Biomass Bioenergy 12 (4) (1997), doi:http://dx.doi.org/10.1016/S0961-9534(97) 00005-6. 241e52.

150. K.O. Davidsson, J.G. Korsgren, J.B.C. Pettersson, U. Jäglid, The effects of fuel washing techniques on alkali release from biomass, Fuel 81 (2) (2002), doi: http://dx.doi.org/10.1016/S0016-2361(01)00132-6. 137e42.

151. M. Daylam-Jafarabad, D. Azadfar, M.H. Arzanesh, The ability to filter heavy metals of lead, copper and zinc in some species of tree and shrub, IJABBR 1 (1) (2013) 53–60.

152. D.B. Sarode, R.N. Jadhav, V.A. Khatik, S.T. Ingle, S.B. Attarde, Extraction and leaching of heavy metals from thermal power plant fly ash and its admixtures, Pol. J. Environ. Stud. 19 (6) (2010) 1325–1330.

153. C. Verwilghen, S. Rio, J. Ramaroson, A. Nzihou, P. Sharrock, The use of hydroxyapatite for the removal of heavy metals from industrial flue gas. PART B: investigation in pilot scale, VERWILGH.pdf.

154. A. Tsangaris, M. Swain, Gas conditioning system, United States, Patent No.: US 8070,863 B2, Date of Patent: Dec. 6, 2011.

155. NOx Control Equipments, Power Engineering, the Magazine of Power Generation for 115 Years (2011), http://www.power-eng.com.

156. E.J. Granite, C.R. Myers, W.P. King, D.C. Stanko, H.W. Pennline, Sorbents for mercury capture from fuel gas with application to gasification systems, Ind. Eng. Chem. Res. 45 (13) (2006)4844–4848, doi:http://dx.doi.org/10.1021/ ie060456a.

157. A. Jain, S.A. Seyed-Reihani, C.C. Fischer, D.J. Couling, G. Ceder, W.H. Green, Ab initio screening of metal sorbents for elemental mercury capture in syngas 1183 streams, Chem. Eng. Sci. 65 (10) (2010)3025–3033, doi:http://dx.doi.org/ 10.1016/j. ces.2010.01.024.

158. M. Ozaki, M.A. Uddin, E. Sasaoka, S. Wu, Temperature programmed decomposition desorption of the mercury species over spent iron-based sorbents for mercury removal from coal derived fuel gas, Fuel 87 (17–18) (2008)3610–3615, doi:http:// dx.doi.org/10.1016/j.fuel.2008.06.011.

159. S. Wu, M. Azharuddin, E. Sasaoka, Characteristics of the removal of mercury vapor in coal derived fuel gas over iron oxide sorbents, Fuel 85 (2) (2006)213–218, doi:http://dx.doi.org/10.1016/j. fuel.2005.01.020.

160. S. Wu, M. Ozaki, M. Uddin, E. Sasaoka, Development of iron-based sorbents for HgO removal from coal derived fuel gas: effect of hydrogen chloride, Fuel 87 (4–5) (2008)467–474, doi:http:// dx.doi.org/10.1016/j.fuel.2007.06.016.

161. A.J. Chandler & Associates, Ltd., review of dioxins and furans from incineration in support of a Canada wide standard review, A Report Prepared for The Dioxins and Furans Incineration Review Group through a Contract Associated with CCME Project 390–2007, December 15, 2006.

162. H.U. Hartenstein, A. Licata, Modern technologies to reduce emissions of dioxins and furans from waste incineration, nawtec08-0010.pdf.

163. F.L. Jones, Process for minimizing dioxin formation during waste and biomass utilization. United States Patent No.: US 2014/0109469 A1, Publication Date: April 24, 2014.

164. http://dutemp.com/plasma_arc/plasma_tech.html   (accessed 08.08.14).

Comparative Analysis of the Production Trial and Numerical Simulations of Gas Production from Multilayer Hydrate Deposits in the Qilian Mountain Permafrost

Youhong Sun[a], Bing Li[a], Wei Guo[a], Xiaoshu Lü[a, b], Yongqin Zhang[a, c], Kuan Li[a, c], Pingkang Wang[d], Guangrong Jin[e], Rui Jia[a], and Lili Qu[a]

[a]College of Construction Engineering, Jilin University, Changchun 130026, PR China

[b]Department of Civil and Structural Engineering, School of Engineering, Aalto University, FIN-02015 Espoo, Finland

[c]Exploration Technology Institute, Chinese Academy of Geological Sciences, Langfang 065000, PR China

[d]Oil and Gas Survey, China Geological Survey, Beijing 100029, PR China

[e]College of Environment and Resources, Jilin University, Changchun 130021, PR China

ABSTRACT

The focus of this paper is to determine the hydrate saturation of the natural gas hydrate (NGH) deposits in the Muri Basin of Qilian Mountain permafrost on the Qinghai–Tibet Plateau and to analyse the performance of different production methods using TOUGH + HYDRATE (T + H) simulator for simulation of the NGH production process trial. Both depressurisation and thermal stimulation methods, using a high-pressure submersible pump to keep the water level in the borehole below the hydrate layer and injecting either hot air or steam, were applied in the hydrate production trial. The simulation of depressurisation-induced gas production under different initial hydrate saturations showed that the gas production was 104.8 m³ for 84 h when the saturation of the hydrate layer was 1.0% and the hydrate abnormal layer was 0.5%. The simulated production was close to the actual value of 102 m³. The result for the NGH deposits indicated a production of 12.8 m³ for the hot air injection-induced gas for 11 h, which is slightly higher than the actual production 10.1 m³. The hot steam injection produced 7.1 m³ of gas for 6 h versus the actual production 3.3 m³. The difference was associated with the simplification of the simulated pressure. The simulation also revealed that larger pressure differences caused the dissociation front of the lower hydrate formation to advance faster than that of the upper formation. Further results also showed that the two heating methods had weak impacts on hydrate dissociation in this production trail since they were used mainly for accelerating the gas migration around the wall of the well, resulting in a slight growth of gas production rate. Depressurisation-induced production was relatively more effective for low saturation hydrate deposits in the Qilian Mountain permafrost.

INTRODUCTION

Natural gas hydrate (NGH) refers to an ice-like cage-type crystalline material that is formed by natural gas and water under low temperature and high pressure conditions, and it is commonly known as "burning ice" (Sloan and Koh, 2007). As a result of its advantages, including large reserve, high energy density, widespread distribution and shallow buried depth, NGH is recognised as one of the most promising and alternative energy options (Collett, 2004 and Makogon, 2010). NGH is mainly found in oceans, lake sediments and terrestrial permafrost formations (Sloan and Koh, 2007). China collected the first physical samples of NGH from the Shenhu area of the South China Sea and Qilian Mountain permafrost in 2007 and 2008 (Zhu et al., 2010a, Zhu et al., 2010b and Zhang et al., 2007). NGH samples were also obtained in the Dongsha area of the South China Sea in 2013 (Ministry of Land and Resources, 2013). To assess the potential NGH production of the Qilian Mountain permafrost, China Geological Survey organised a production field test in 2011. Both depressurisation and thermal stimulation methods were used in this test, which produced a total of 101 m^3 of natural gas for 95 h (Li, 2012).

One of the main purposes of NGH research is to achieve commercial production (Sun et al., 2014). There are currently three broad gas production schemes according to the production principles. The first dissociates NGH through breaking down NGH's phase equilibrium by changing the temperature, pressure and salinity conditions using techniques of thermal stimulation, depressurisation and chemical inhibitor injection (Sun et al., 2014, Cheng et al., 2013, Moridis et al., 2011, Li et al., 2012 and Li et al., 2007). The second scheme is based on replacement of guest molecule, in which one or more types of gases are injected into the NGH deposit to extract the methane (CH_4), such as the commonly-used carbon dioxide (CO_2) replacement method (Ebinuma, 1993; Ohgaki et al., 1996). The third scheme is similar to those used to recover solid mineral resource. The samples containing solid hydrates are recovered and brought to the surface to dissociate and produce natural gas, which mainly occurs through methods such as the hydraulic lifting and the assembled injection pump methods (Fan, 2012 and He, 2013). Presently, depressurisation and thermal stimulation methods are the most commonly-used schemes in NGH production trials (Sun et al., 2014, Makogon and Omelchenko, 2013, Hancock et al., 2005, Dallimore et al., 2008, Japan Oil Gas and Metals National, 2013 and Jia, 2013). Numerical simulation is also widely applied to predict the dynamic properties of gas hydrate production. Zhao et al. (2013) and Li (2012) used TOUGH + HYDRATE to predict the production potential of hydrate deposits in the Qilian Mountain permafrost. Both depressurisation in a single vertical well and huff and puff method in a single horizontal well were simulated. However, their simulations were conducted for one thick hydrate-bearing layer, which cannot be applied to the gas hydrate deposit in this research.

This paper describes the simulation results from the NGH project of production trial conducted in the Qilian Mountain permafrost in 2011 as a case study. TOUGH + HYDRATE was used to simulate the production and results were analysed and compared against the actual production data for studying the formation properties and the performance of different production methods.

NGH PRODUCTION TRIAL PROJECT

Geological Background

The well DK-8 of NGH production trial is located in Muri, Tianjun County of Qinghai Province and next to the Juhugeng mine of the Muri coalfield in the Muri Basin in the Qilian Mountain permafrost, as shown in Fig. 1. Qilian Mountain permafrost is located in the north-eastern margin of the Tibetan Plateau, which has the permafrost area about 10 × 10⁴ km². The permafrost zone is dominated by alpine permafrost with seasonal frozen soils in the valleys and foothills (Zhou et al., 2000).

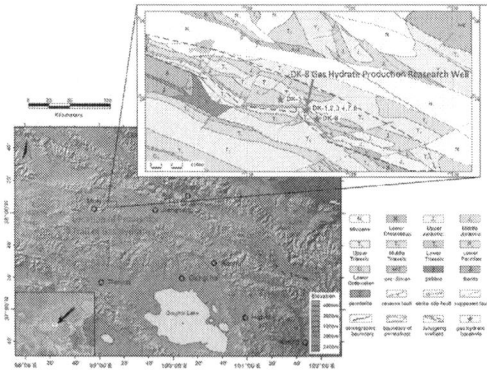

Figure 1: Location of the test area of NGH production trial project in Qilian Mountain permafrost (Wang et al., 2014).

The altitude of the test area is 4070 m with the average ground temperature −2.4 °C, and the average permafrost thickness 95 m (Zhou et al., 2000 and Wang et al., 2014). The permafrost provided suitable conditions (pressure and temperature) for the formation of NGH (Zhu et al., 2006). Tectonically, Qilian Mountain is divided into Southern, Central and Northern Qilian. The test area is located in the rift-depression basin between the tectonic units of Central Qilian and Southern Qilian. The rift basin was formed during the reactivation of the deep fault system of Northern Qilian during Yanshanian (Lu et al., 2010a and Lu et al., 2010b), and the fault system provided channels and reservoir space for transporting gas and NGH formation. The following four sets of source rocks are found in the Muri basin: Carboniferous dark mudstone (limestone), Lower Permian dark limestone, Upper Triassic Galedesi Formation dark mudstone and Jurassic dark shale. These high-quality source rocks are in the mature – over mature stage and have a strong ability to generate gas (Fu and Zhou, 1998). The geological condition provided three prerequisites for NGH formation: overburden and temperature/pressure conditions, reservoir space/transport channels, and gas-generating source rocks.

NGH Geological Features

The China Geological Survey conducted 10 NGH scientific drillings in the South Syncline of the Juhugeng mine during 2008–2013 and collected the physical samples from the wells of DK-1, DK-2, DK-3, DK-7, DK-8 and DK-9. A series of abnormal NGH-associated signs including high-pressure gas, wellhead kick, presence of water seepage and low temperatures of the crack surface were observed in the wells of DK-4, DK-5, DK-6 and DK-10 during the drilling (Wang et al., 2014 and Ministry of Land and Resources, 2013).

The field observations indicated that two types of NGH occur in the cores: fracture-filling and pore-space NGHs. The former fills the fractures of siltstone, mudstone and oil shale as lamellar flakes and agglomerates a few millimetres thick layer of gas hydrate; they are the main type of occurrence and visually observable. Another type of NGHs is disseminated in sandstone pore which is difficult to observe visually; but it can be indirectly inferred from the features such as dispersive low-temperature abnormalities in infrared temperature, bubbles or droplets continually emerging from the core (Zhu et al., 2010a, Zhu et al., 2010b and Wang et al., 2013). NGHs mainly occur in the Middle Jurassic Jiangcang Formation beneath the permafrost at depths between 133.0 and 396.0 m (Wang et al., 2011). The deposit lithology has generally low permeability tight rock, which is not favourable for NGH production.

Laboratory tests showed that the gas components of NGHs from Qilian Mountain were complex. In addition to methane, there were propane and ethane, and some samples contained small amounts of CO_2 also. A hydrocarbon isotope analysis showed clear features of deep thermogenic gas rather than shallow microbial gas (Zhu et al., 2010a and Zhu et al., 2010b).

Production Well Design

The depressurisation (through lowering the water level) and thermal (through injecting hot air and steam) simulations in a single vertical well were selected in the production trial. The well was designed using a pilot hole drilled to obtain the core in order to determine the location of the NGH layers. The hydrates were mainly within a range of 130–330 m interval as shown in Fig. 2. After the NGH layers were located, a 215 tricone bit was used for a borehole at a depth of 330 m, and then another 156 roller bit continued all the way down to the bottom. The 168 casing was placed into the hole at the depth of 330 m. The well screens were employed at the NGH and NGH abnormal layers in place of conventional casings, for a total of 10 segments (Fig. 2). Considering the fracture-filling structure that dominates the occurring NGH, the length and location of the well screens were determined based on the fracture angle and desired production radius.

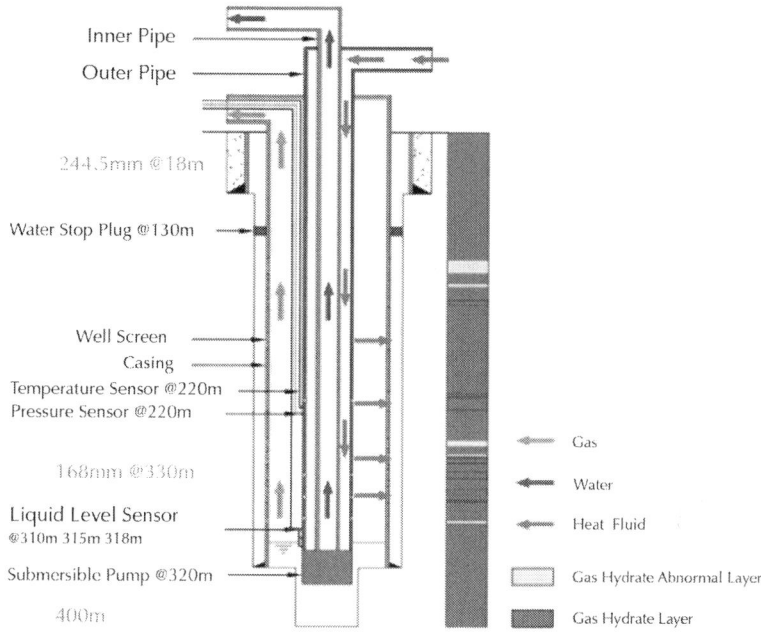

Figure 2: NGH layer and NGH abnormal layer in the DK-8 well, and the production trial well design.

As shown in Fig. 2, the double-wall drill pipe attaches the submersible pump and level, temperature and pressure sensors. The submersible pump and level sensors are used for depressurisation-induced production, and the temperature and pressure sensors are installed inside the borehole to conduct the real time monitor of the dissociation of NGHs.

Production Process and Performance

The depressurisation-induced method was initially used in the production trial by controlling the water level in the borehole with a high-pressure submersible pump fixed to the bottom of the double-wall drill pipe. Its outlet was connected to the inner pipe of the double-wall drill pipe, and water in the borehole was discharged to the surface through the inner pipe (marked with blue arrow (in the web version) in Fig. 2). The outlet of pump was located at a depth of 320 m. Sensors at three levels (upper 2 m, middle 5 m and lower 10 m above the outlet of the pump), were installed. A level controller controlled the pump based on the signal from the sensors. The water level could be kept below the production layer through the water level sensors (at a depth between 310 and 315 m). The production pressure was assumed to atmospheric pressure because the NGH layer around the wellhead exposed to atmosphere. The lower level sensor was a failsafe. The measured production of depressurisation-induced method was 81.97 m³ for 84 h.

The hot air injection method was for validating the performance of the thermal stimulation-induced production. The average temperature of injected air, heated by electromagnetic or solar heat, was approximately 50 °C. The production with this method was 9.73 m³ for 11 h.

The steam injection method was used in thermal stimulation-induced production, with an average temperature of injected steam of 150 °C. The heating time was 5.3 h, including the injection time and well-closing time, and the production lasted for 42 min. The entire process took 6.0 h and produced 3.3 m³ of gas.

NUMERICAL MODEL

Numerical Simulator

The numerical simulations were conducted using the TOUGH + HYDRATE v1.0 simulator (Moridis et al., 2008). TOUGH + HYDRATE v1.0, developed by Moridis at the Lawrence Berkeley National Laboratory, USA, is widely used in gas hydrate simulation. Grover used this model to analyse the observed production data from the Messoyakha field (Grover et al., 2008), and Moridis and Li used it to simulate depressurisation-induced gas production from class 1 hydrate deposits and in a pilot-scale hydrate simulator (Moridis et al., 2007 and Li et al., 2014).

Assumptions and Initial Conditions

The following assumptions were used for the numerical simulation:

- The salinity was assumed to be zero because it was too low to affect the temperature and pressure phase equilibrium condition of NGHs and liquid phase temperature of water (Collett, 2004).

- Methane was assumed to be the only gas component because most of the hydrocarbon gas is CH_4 (Zhao et al., 2013, Li et al., 2012 and Moridis et al., 2008).

The geothermal gradients were assumed to be constant within and below the frozen layer. The values were determined based on the log data and literature (Jin et al., 2011 and Chen et al., 2005), as listed in Table 1. Then the formation temperature distribution can be obtained using the following equations:

$$T = T_0 + G_p z \quad z \leq H_s \tag{1}$$

$$T = T_s + G_{bp}(z - H_s) \quad z > H_s \tag{2}$$

where T_0 is the ground surface temperature, Gp and Gbp are the geothermal gradients of the permafrost and below permafrost, respectively, z is the depth, Hs is the depth of the permafrost base and Ts is the temperature of solidus.

Table 1: Physical parameters for gas hydrate deposits at the DK-8 drilling site

Properties	NGH layer	NGH abnormal layer	Non-NGH layer
Porosity	0.07	0.05	0.03
Intrinsic permeability	4.00e-15 m²	2.00e-15 m²	1.00e-15 m²
Average density (all formations)	2450 kg/m³		
Thermal gradient below the frozen layer	0.03 K/m		

Thermal gradient within the frozen layer	0.0182 K/m
Frozen layer thickness	110 m
Permafrost ground temperature	−2.4 °C
Initial temperature (at base of permafrost)	−0.4 °C
Initial pressure (at base of permafrost)	2.643795
Thermal conductivity at full saturation	2.4 W/(m K)
Thermal conductivity of dry rocks	0.5 W/(m K)
Composite thermal conductivity model	$K_\theta = K_{dry} + (\sqrt{S_A} + \sqrt{S_H}) \cdot (K_{wet} - K_{dry}) + \alpha S_1 K_1$
Capillary pressure model	$P_{cap} = -P_{01}[(S^*)^{-1/} - 1]^{1-}$ $S^* = (SA - S_{irA})/(S_{mxA} - S_{irA})$
SirA	0.29
	0.45
P_{01}	10^5
Relative permeability model	$K_{rA} = n(S^*)$ $K_{rG} = (S_G^*)^{nG}$ $S_A^* = (S_A - S_{irA})/(1 - S_{irA})$ $S_G^* = (S_G - S_{irG})/(1 - S_{irG})$ EPM #2 model
N	3.572
nG	3.572
SirA	0.05
SirG	0.25

The permafrost thaws and freezes when the temperature is within the thawing range between the solidus and liquidus temperatures depending on the pressure, grain size and salinity of the pore water

(Nixon, 1986 and Konrad and Seto, 1991). The solidus temperatures are 0.2 °C lower than the liquidus temperatures (Galushkin, 1997). Because the salinity in the water was ignored, the liquidus temperature at the base of the Muri Basin was determined as follows:

$$T_L = T_S + 0.2 = 0 - 0.0731P$$

(3)

We assumed that the rock matrix porosity and fractures were fully occupied by water at the temperatures above TL and ice at temperatures below Ts. The pressures in the permafrost and sediments below were determined based on the lithostatic and hydrostatic pressure, respectively:

$$P = P_0 + \rho_p g z 10^{-6} \quad z \leq H$$

(4)

$$P = P_s + \rho_w g(z - H)10^{-6} \quad z > H$$

(5)

where P_0 is the atmospheric pressure at the ground surface, p and w are the densities of permafrost and water, respectively, g is the gravity acceleration constant (9.81 m/s^2) and Ps is the pressure of the permafrost base at temperature Ts.

By solving Equations (1), (3) and (4), the temperature and pressure of the permafrost base were calculated as −0.4 °C and 2.643795 MPa, respectively, and the position of the permafrost base was at 110 m, which was similar to the observed values. The temperature and pressure below the base of the model can be calculated using Equations (2) and (5).

Geometry, Domain Discretisation, and System Properties

The type of NGH deposits near the well must be determined to set up the model. Based on the previous analysis of NGH characteristics, the deposits were multilayer and considered to be of class 3, because no gas and water were observed below the NGH deposits. Although intervals with abnormally high gas pressure above the stable NGH layers were found in some boreholes, we considered that these gases could not flow into the NGH layer because of the impermeable rock. A 2-D model with NGH layer and NGH abnormal layer simplification was applied. There were 5 NGH layers (171.5–175 m; 228–237 m; 266–269 m; 273.5–278 m; and 287.5–292 m) and 3 NGH abnormal layers (148–160 m; 253–259 m; and 302.5–305 m). Therefore, the region from 130 to 330 m was selected for simulation, as shown in Fig. 3. The source/sink settings for the NGH and NGH abnormal layers are shown in Fig. 3. The radius of the impact was small because production time was short. Therefore, the radius of the model was set to be 50 m.

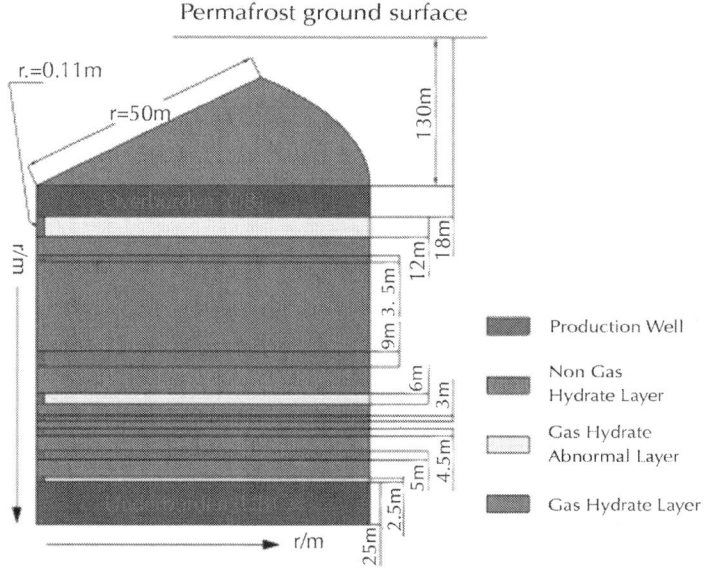

Figure 3: Numerical model of the hydrate deposit and production well of the DK-8 well.

The radial grid model was discretized into 91 cells in the R-direction, and the Z-direction was 109 layers. The uppermost and lowermost elements were set to the inactive boundaries, through which heat and mass can be transferred to the simulation domain. The pressures and temperatures of the inactive boundaries remain constant over time. The observation of the core that was recovered from the test well showed that the rock was tight with low in porosity; therefore, the non-NGH layer was treated as low permeability layer in the model. The NGH abnormal layer had relatively high porosity and contained a small number of pore-space hydrates, the NGH layers contained a small amount of fractures, and the hydrate saturation was relatively high. The physical properties and initial conditions of the model, based on the core observations, laboratory test analyses and relevant literature (Zhao et al., 2013, Li et al., 2012, Collett and Dallimore, 2003, Moridis et al., 2005, Jin et al., 2011 and Chen et al., 2005), are shown in Table 1.

Production Parameter

The production process was divided into three phases. The first phase included depressurisation-induced production and the production pressure for all layers was set as atmospheric pressure (0.06 MPa) based on the reported values and measurements. The second phase included the hot air injection-induced production. The heating power was calculated at 134 W based on the temperature and flow rate of the injected air as well as the strata temperature and borehole structure. The third phase included the steam injection-induced production. The water heating capacity of the boiler was 54.7 kg/h (Jia, 2013); if 60% of the steam was injected into the planned producing strata (after the loss along the way) with the injection flow rate 2×10^{-4} kg/s per metre of the producing strata. The average temperature of the injected steam was 147 °C (Li, 2012) with enthalpy 2742.66 kJ/kg. The water level was maintained below the NGH layer during the heat injection process to ensure that the depressurisation-induced production occurred simultaneously.

SIMULATION AND RESULTS ANALYSIS

Determining the NGH Saturation of the Reservoir

Currently, there are the limited numbers of published data on the NGH saturation of the Muri Basin. Lu reported a hydrate saturation of 5–10%

(Lu et al., 2010a and Lu et al., 2010b). Based on the field observations and studies of Wang and Zhu (Zhu et al., 2010a, Zhu et al., 2010b and Wang et al., 2013), we assumed that the NGH abnormal layer was dominated by pore-space NGH at a low saturation, whereas the NGH layer contains both fracture-filling NGHs and pore-space NGHs at relatively high saturation. Because the other basic parameters were fixed, we simulated the depressurisation-induced production process at different saturations to determine the NGH saturation.

The simulation showed that when the saturations of the NGH layers and the NGH abnormal layers were 5%, 3%, 1.5%, 1%, 0.5% and 0.25%, and 2.5%, 1.5%, 0.5%, 0.5%, 0.2% and 0.1%, the corresponding volumetric flow rates of methane production were 108.8, 54.2, 29.7, 27.6, 26.3 and 25.7 m^3/day, and the cumulative volumes of methane were 465.3 m^3, 227.4 m^3, 115.5 m^3, 104.8 m^3, 97.2 m^3 and 93.1 m^3, respectively. Fig. 4 shows the volumetric flow rate of methane production at the different saturations indicating that the production flow rate decreased as gas hydrate saturation decreased. Similar pattern of cumulative methane production can be observed in Fig. 5. Fig. 6 shows that the radius of the dissociation gradually increased with gas hydrate saturation decreasing. That is because hydrate decomposition is considered as the most important component of heat consumption in depressurisation-induced production and hydrate saturations have a great impact on hydrate dissociation progress. Therefore, the production rate and cumulative gas production did not decrease proportionally to the saturation levels. However, the amount of gas generated during the actual construction (approximately 20 m^3), e.g., drilling, equipment commissioning and process interruptions, was not recorded, so the total gas production was 102 m^3because the recorded production volume was 82 m^3. This number was similar to the production volume corresponding to a hydrated saturation of 1.0%. Therefore, the NGH saturation around the well was 1.0%, and the saturation of the NGH abnormal layer was 0.5%.

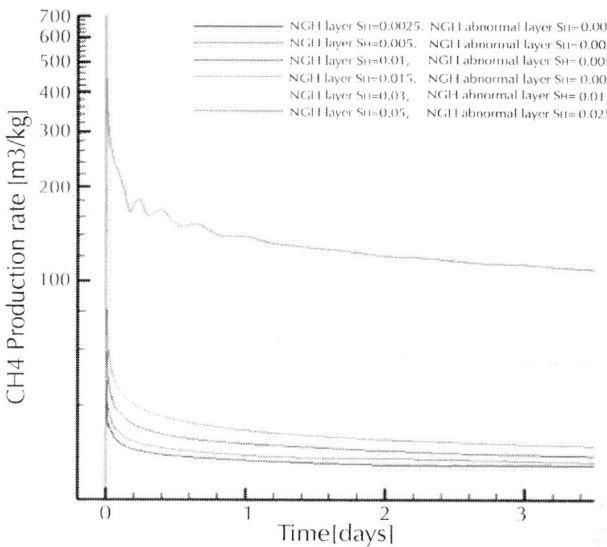

Figure 4: Volumetric gas flow rate of methane production for six different hydrate initial saturation cases.

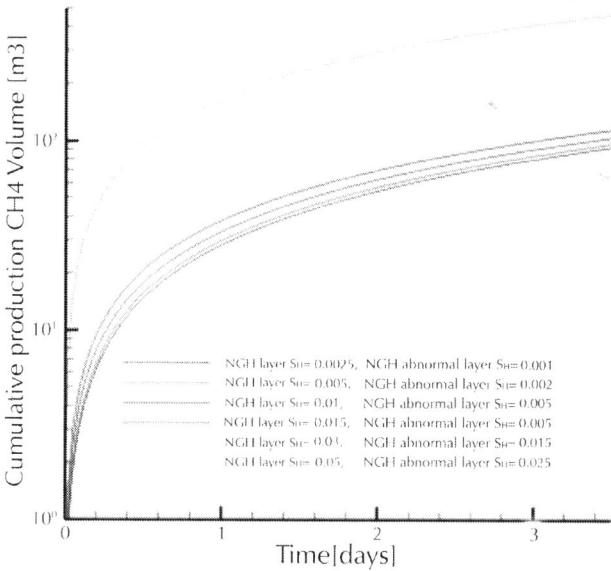

Figure 5: Cumulative methane production volume for six different hydrate initial saturation cases.

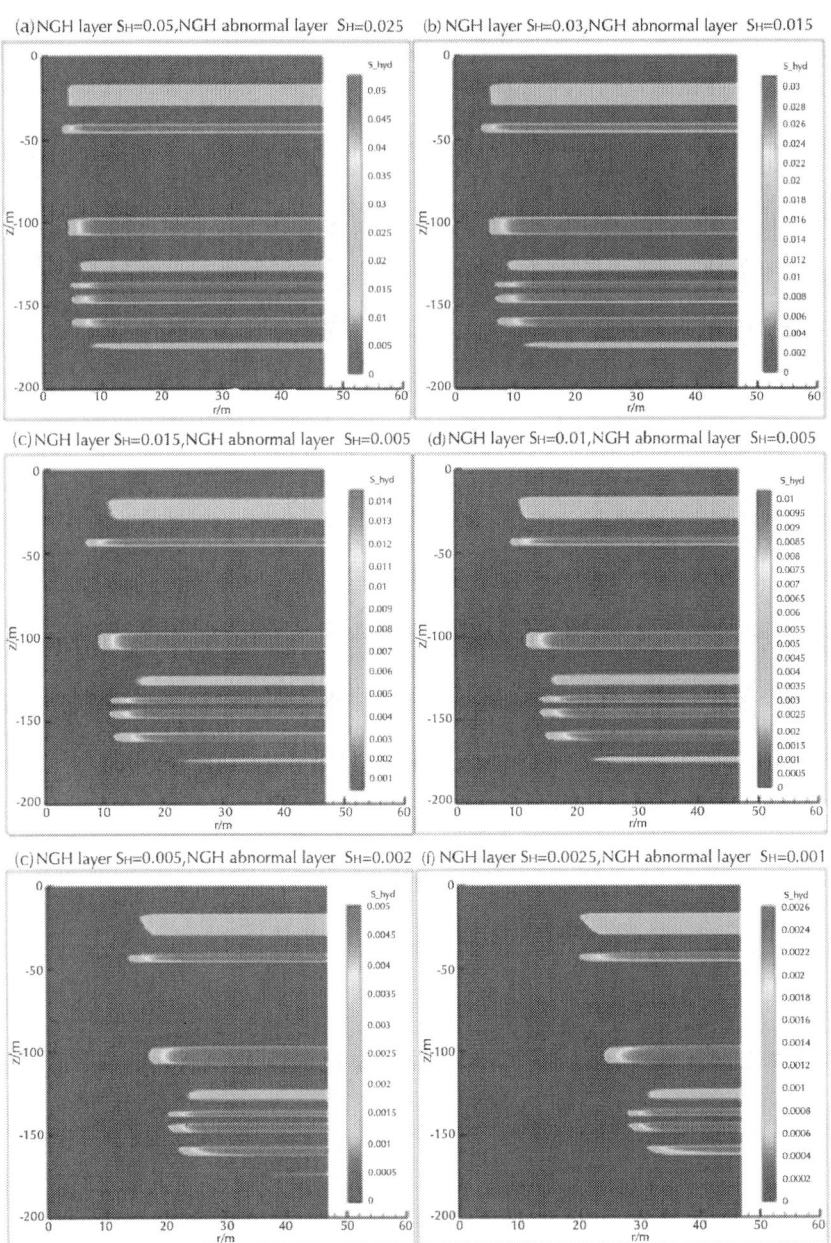

Figure 6: Hydrate saturation distribution after depressurisation-induced production for six different initial saturations.

Depressurisation-induced Production

NGH production can be analysed when initial NGH saturation has been determined. First, we discussed the results of the simulated depressurisation-induced production. The blue line (in the web version) inFig. 4 shows that the gas flow rate increased rapidly, reaching a maximum of 76.8 m³/day, and then decreased with the rate of decline slowing and starting to stabilise after 10 h of production. This phenomenon occurred for the following reasons: (1) the gas from NGHs near the well could quickly enter the well unit after dissociation, resulting in a high production rate; (2) the dissociated gas required more time to reach the well as the hydrate dissociation front moved forward, resulting in a decreased well production rate; (3) the driving force of dissociation decreased as additional NGH-dissociated gas accumulated in the deposit, which caused the gas pressure at the dissociation front to increase and results in a decreased dissociation rate and decreased well production rate; (4) NGH dissociation is an endothermic reaction, which leaded to a decrease in both the temperature of the surrounding formation and the production rate; and (5) as the dissociation front advanced further, the impacts of previously mentioned factors on the NGH decomposition rate decreased, and the hydrate production rate declined more slowly and eventually stabilised.

Fig. 7 shows the spatial distribution of hydrate saturation at 15 min at days 1, 2 and 3, which shows that lower hydrate saturation caused a faster advance of the dissociation front due to the relatively high gas–liquid permeability. In addition, the hydrate dissociation front advanced at a greater speed as the location of the hydrate layers increased in depth due to the increased formation pressure; The formation-well pressure difference, therefore, increased, and the driving force of hydrate dissociation became greater. For single-layer hydrates, the hydrate dissociation front advanced in parallel to the NGH layer at the beginning of production (shown in Fig. 7 a–b) and this phenomenon became more significant in thicker NGH layers. As the NGH dissociation front continued to advance, the temperature of the dissociation front decreased, in addition, the temperature of the NGH layer boundaries decreased more slowly because of the non-NGH layer, which resulted in a faster dissociation rate of the NGH layer boundaries and an arc-shaped dissociation front. The dissociation rate at the NGH layer base was faster because of the higher pressure and temperature differences at the base, as shown in Fig. 7 c–d.

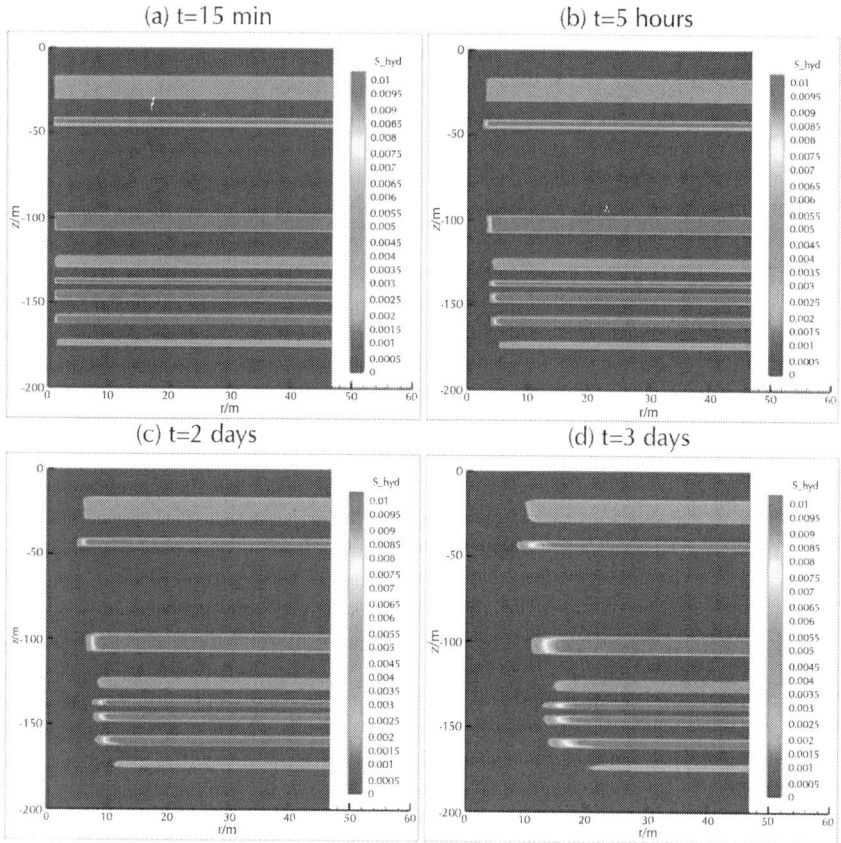

Figure 7: Hydrate saturation distributions at the end of 15 min, 1 day, 2 days, and 3 days (depressurisation-induced production).

Hot Air Injection-induced Production

Fig. 8 shows the gas production rate and cumulative gas production using the heat injection method and depressurisation method. The production went from 104.8 to 117.6 m³ with increased production 12.8 m³, slightly greater than the actual production of 10.1 m³. In addition, the gas production rate increased initially and then fell slightly after five hours. Both the NGH saturation and the temperature distributions should be considered in interpreting this phenomenon. Fig. 9 displays the formation temperature distribution after 11 h of production indicating that the impact range of the heating was approximately 0.5 m. Fig. 10illustrates the spatial distributions of hydrate saturation at 15 min and 11 h showing that the hydrate dissociation front moved 2 m forward from 10 to 22 m, which was much larger than the impact range of the temperature. Therefore, it was believed that heating the formation using hot air injection had little effect on hydrate dissociation during the production trial. Because of the changes in the distribution of gas saturation, the early increase of gas production rate in the heat injection-induced production was considered a result of the temperature rise, which caused faster gas migration in the near-well deposits. Because of the small impact range of temperature, the gas production rate began to fall as the gas saturation in the near-well deposits decreased.

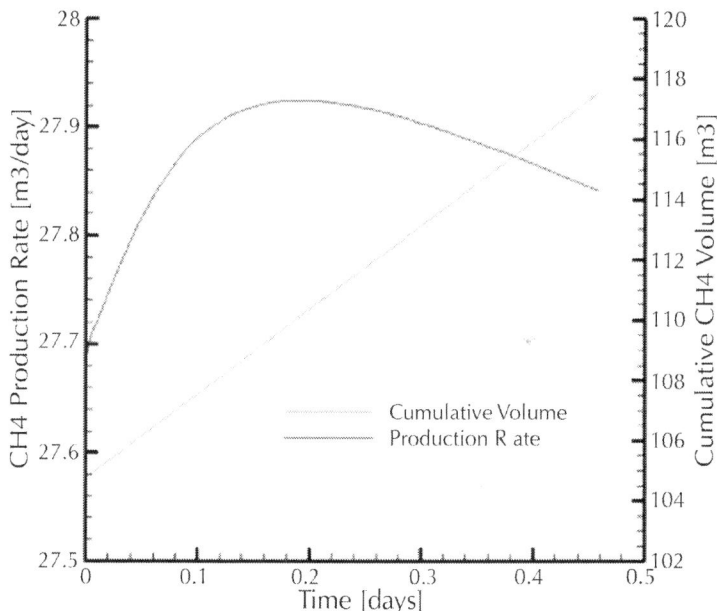

Figure 8: Volumetric methane production rate and cumulative gas production (hot air injection-induced production).

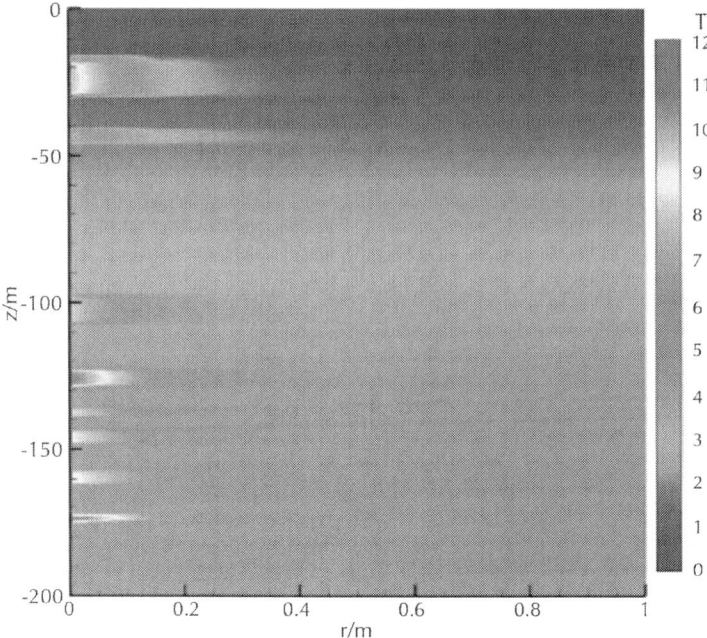

Figure 9: Temperature distribution after 11 h of hot air injection-induced production.

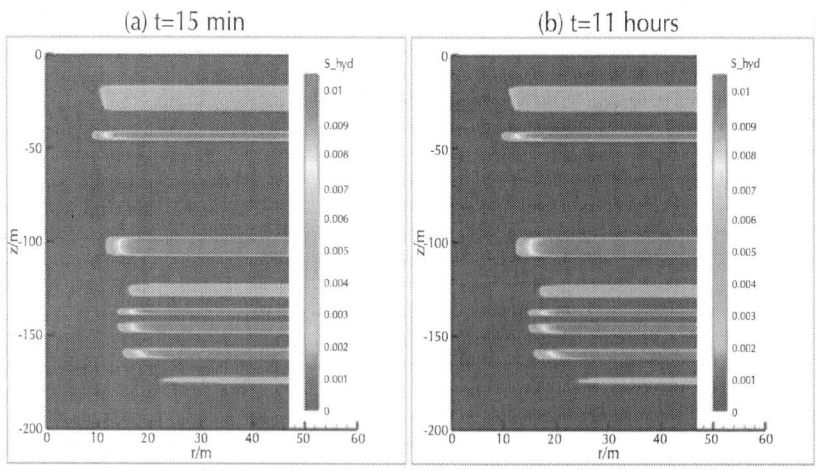

Figure 10: Hydrate saturation distribution at the end of 15 min, 11 h (hot air injection-induced production).

Steam Injection-induced Production

The production process can be divided into the following two phases: the combined phase of steam injection and depressurisation and the depressurisation phase. Fig. 11 shows the gas production rate and cumulative gas production during the production process indicating that the total production increased from 117.6 m^3 to 123.8 m^3 (eventually to 124.7 m^3), and a total of 7.1 m^3 was produced, which was slightly more than the actual production of 3.3 m^3. We believed that the pressure inside the borehole increased slightly after the production. The pressure increase was ignored in the simulation, which resulted in a slightly greater production.

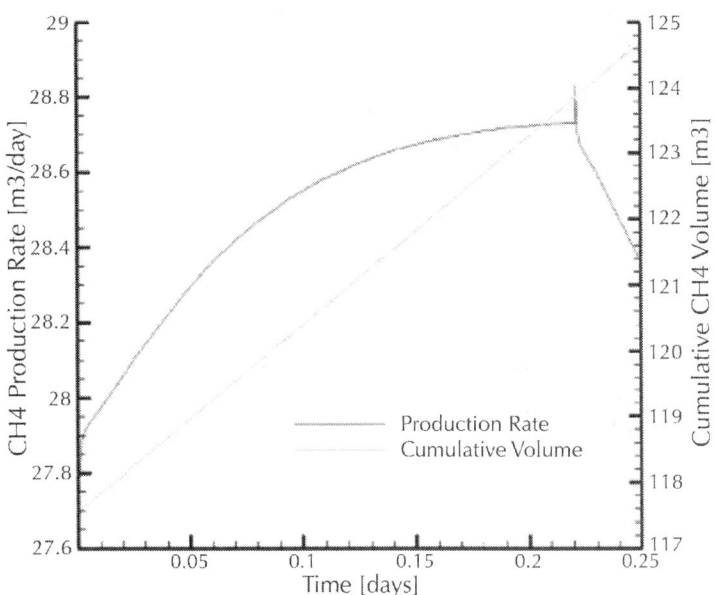

Figure 11: Volumetric methane production rate and cumulative gas production (steam injection-induced production).

Fig. 11 shows that the gas production rate continued to increase during the heat injection process; however, the increase gradually slowed down, so the overall change was small. This phenomenon also can be explained by NGH's saturation and formation temperature distributions. Fig. 12 illustrates the formation temperature distribution after 5.3 h of heating-induced production showing that the impact range of heating was up to 0.6 m. Fig. 13 shows the spatial distribution of NGH saturation at 15 min and 6 h indicating that the hydrate dissociation front advanced 1 m from its position in the last production step. Therefore, heating the formation with steam injection had no effect on hydrate dissociation. This conclusion also can be drawn from the water saturation distribution. The impact range of the injected steam did not reach the hydrate dissociation front, and the gas production continued to increase and the magnitude of the increase was greater than that in the hot air injection-induced production because of the larger impact range in the heated formation and higher formation temperature (up to 31.4 °C). Therefore, the gas migrated more rapidly and gas saturation of the near-well deposits decreased, resulting in a slower increase of the gas production rate.

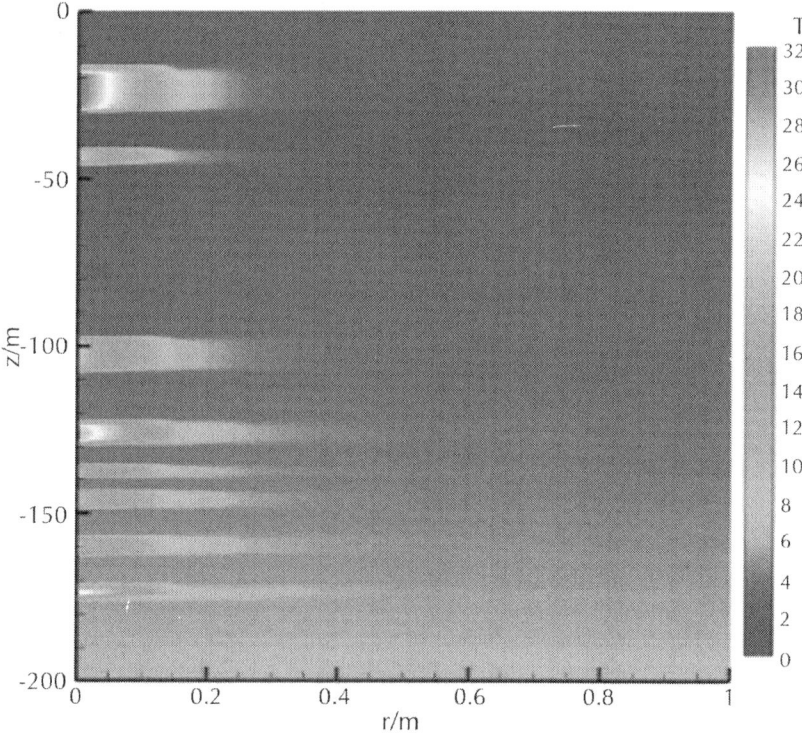

Figure 12: Temperature distribution after 5.3 h of steam injection-induced production.

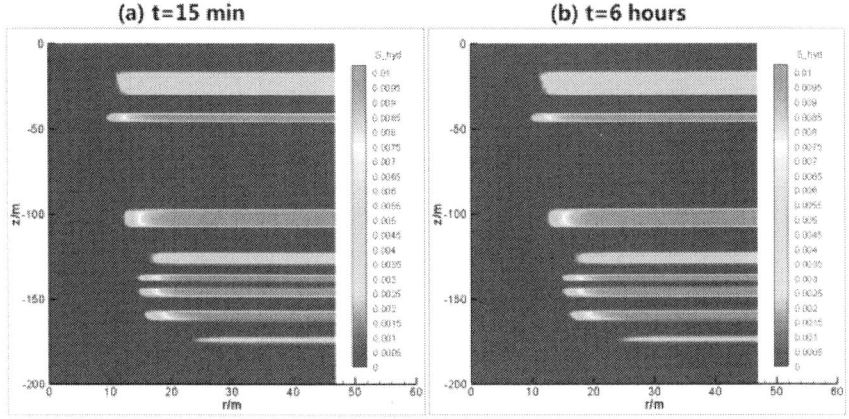

Figure 13: Hydrate saturation distribution at the end of 15 min, 6 h (steam injection-induced production).

As shown in Fig. 11, the gas will move more rapidly toward the wellhead after the hot steam injection ceased because fluids were not injected into the formation. Therefore, the gas production rate increased rapidly and then decreased quickly because of the small gas volume. The gas production rate decreased due to the formation temperature drops without an external heat supply.

CONCLUSIONS

The production parameters of the production trial project in DK8 well were investigated; the process was simulated to analyse the patterns of well production, hydrate saturation and temperature distribution. The following conclusions can be drawn:

- At the relatively low gas hydrate saturation, the production rate and cumulative gas production did not decrease proportionally to the decrease of gas hydrate saturation. That is because the hydrate dissociation radius will evidently increase.

- The hydrate saturations of the NGH layer and NGH abnormal layer were 1% and 0.5%, respectively, by process simulations of depressurisation-induced production for different initial saturations. This saturation level is lower than that estimated by Lu, which was reported to be 5–10%.

- The average gas production was 30 m³/day for 3.5 days of depressurisation-induced production, with the gas production gradually decreasing to 27.7 m³/day. During the 11 h of hot air injection-induced production, the gas production rose from 27.7 to 27.9 m³/day and then dropped to 27.8 m³/day. For the 5.3 h of steam injection-induced production, the gas production went up from 27.8 to 28.7 m³/day. The depressurisation-induced gas production fell to 28.4 m³/day within 0.7 h after the heat injection stopped.

- Because of limited heat injection time, the heat-affected ranges of the two heat injections were 0.5 m and 0.6 m, respectively, which had no effect on the hydrate dissociation but accelerated the near-well liquid/gas migration and caused small fluctuations in gas production. In addition, the pulse injection will accelerate production gas migration.

- The simulation results in this production trial showed that the steam injection method is more capable of heating the formation than the hot air injection method; however, steam injection requires larger equipment and greater power consumption. Overall, the heating methods are not suitable for recovering formations with low hydrate saturation in the Muri area of the Qilian Mountains because heating formation is a slow process. However, the hydrate dissociates quickly at low saturation Therefore, the depressurisation method is still the best choice for the next production trial project.

ACKNOWLEDGEMENTS

This study has been supported by China Geological Survey Project (GZHL201400307, GZHL20110326), and National Natural Science Foundation of China (Grant No. 51474112, Grant No. 51304079, Grant No.41102021).

REFERENCES

1.   Chen, D.F., Wang, M.C., Xia, B., 2005. Formation condition and distribution prediction of gas hydrate in QinghaieTibet plateau permafrost. Chin. J. Geophys. 48 (1), 179e187.

2. Cheng, Y.F., Li, L.D., Yuan, Z., Wu, L.Y., Sajid, M., 2013. Finite element simulation for fluidesolid coupling effect on depressurization-induced gas production from gas hydrate reservoirs. J. Nat. Gas Sci. Eng. 10, 1e7.

3. Collett, T.S., 2004. Gas hydrates as a future energy resource. Geotimes 49, 24e27.

4. Collett, T.S., Dallimore, S.R., 2003. Permafrost-associated Gas Hydrate in Natural Gas Hydrate. Springer Netherlands, pp. 43e60.

5. Dallimore, S.R., Wright, J.F., Nixon, F.M., 2008. Geologic and porous media factors affecting the 2007 production response characteristics of the JOGMEC/NRCAN/ AURORA Mallik gas hydrate production research well. In: Proceedings of the 6th International Conference on Gas Hydrates. Vancouver, British Columbia, Canada; July 6e10. http://www.isope.org.

6. Ebinuma, T, Nov 16 1993. Method for dumping and disposing of carbon dioxide gas and apparatus therefor. United States Patent US 5261490.

7. Fan, F.C., Mar 14 2012. New method for exploiting gas hydrate. China Patent CN 102373909A.

8. Fu, J.H., Zhou, L.F., 1998. Carboniferous-Jurassic stratigraphic provinces of the southern Qilian basin and their petro-geological features. Northwest Geosci. 19 (2), 47e54.

9. Galushkin, Y., 1997. Numerical simulation of permafrost evolution as a part of sedimentary basin modeling: permafrost in the PlioceneeHolocene climate history of the Urengoy field in the West Siberian basin. Can. J. Earth Sci. 34 (7), 935e948.

10. Grover, T., Moridis, G.J., Holditch, S.A., July 06e11, 2008. Analysis of reservoir performance of Messoyakha gas hydrate field. In: Proceedings of the Eighteenth International Offshore & Polar Engineering Conference. Vancouver, British Columbia, Canada.

11. Hancock, S.H., Collett, T.S., Dallimore, S.R., Satoh, T., Inoue, T., Huenges, E., Henninges, J., Weatherill, B., 2005. Overview of thermal-stimulation production-test results for the JAPEX/JNOC/ GSC et al. Mallik 5L-38 gas hydrate production research well. Bull. Geol. Surv. Can. 585, 15 p.

12. He, Y.F., June 19 2013. Marine gas hydrate electronic-spraying pump composite exploitation method and apparatus. China Patent CN 102392646B.

13. Japan Oil, Gas and Metals National Corporation [Internet]. Tokyo; c2014 [updated 2013 Mar 18; cited 2014 Feb 12]. Flow Test from Methane Hydrate Layers Ends. [about two screens] Available from: http://www.jogmec.go.jp/english/news/ release/ news_01_000005.html.

14. Jia, R., 2013. Research and Experiment on Steam Mining System and Heat Pipe Rapid Freezing Mechanism for Gas Hydrate. Jilin University, Chang Chun.

15. Jin, C.S., Qiao, D.W., Lu, Z.Q., Zhu, Y.H., 2011. Study on the characteristics of gas hydrate stability zone in the Muri permafrost, Qinghai-comparison between the modeling and drilling results. Diqiu Wuli Xuebao Freezing of a clayey silt contaminated within organic solvent. J. Contam. Hydrol. 8, 335e355.54 (1), 173e181.

16. Konrad, J.M., Seto, J.T.C., 1991.

17. Li, B., Li, X.S., Li, G., Feng, J.C., Wang, Y., 2014. Depressurization induced gas production from hydrate deposits with low gas saturation in a pilot-scale hydrate simulator. Appl. Energy 129, 274e286.

18. Li, G., Li, X., Tang, G.L., Zhang, Y., 2007. Experimental investigation of production behavior of methane hydrate under ethylene glycol stimulation in unconsolidated sediment. Energy Fuel 21 (6), 3388e3393.

19. Li, K., 2012. Simulation and Field Experience of Steam Mining System for Permafrost Gas Hydrate. Jilin University, Chang Chun.

20. Li, X.S., Li, B., Li, G., Yang, B., 2012. Numerical simulation of gas production potential from permafrost hydrate deposits by huff and puff method in a single horizontal well in Qilian Mountain, Qinghai province. Energy 40 (1), 59e75.

21. Lu, Z.Q., Zhu, Y.H., Zhang, Y.Q., Wen, H.J., Li, Y.H., Jia, Z.Y., Liu, C.L., Wang, P.K., Li, Q.H., 2010a. Basic geological characteristics of gas hydrates in Qilian Mountain permafrost area, Qinghai Province. Miner. Depos. 29 (1), 182e191.

22. Lu, Z.Q., Zhu, Y.H., Zhang, Y.Q., Wen, H.J., Li, Y.H., Wang, P.K., 2010b. Estimation method of gas hydrate resource in the Qilian

Mountain permafrost area, Qinghai, Chinaea case of the drilling area. Geol. Bull. China 29 (9), 1310e1318.

23. Makogon, Y.F., 2010. Natural gas hydratesea promising source of energy. J. Nat. Gas Sci. Eng. 2 (1), 49e59.

24. Makogon, Y.F., Omelchenko, R.Y., 2013. Commercial gas production from Messoyakha deposit in hydrate conditions. J. Nat. Gas Sci. Eng. 11, 1e6.

25. Ministry of Land and Resources, PRC [Internet]. Beijing; c1999e2007 [updated 2013 Dec 17; cited 2014 Feb 12]. Press conference about exploration achievements of gas hydrate in the sea areas, MLR 2013. [about two screens]. Available from: http://www.mlr.gov.cn/wszb/2013/trq/.

26. Moridis, G.J., Kowalsky, M.B., Pruess, K., 2008. TOUGHþHYDRATE v1.0 User's Manual: a Code for the Simulation of System Behavior in Hydrate-bearing Geologic Media. Lawrence Berkeley National Laboratory, Berkeley, CA, USA.

27. Moridis, G.J., Kowalsky, M.B., Pruess, K., 2007. Depressurization-induced gas production from class-1 hydrate deposits. SPE Reserv. Eval. Eng. 10 (05), 458e481.

28. Moridis, G.J., Reagan, M.T., Boyle, K.L., Zhang, K.N., 2011. Evaluation of the gas production potential of some particularly challenging types of oceanic hydrate deposits. Transp. Porous Media 90 (1), 269e299.

29. Moridis, G.J., Seol, Y., Kneafsey, T.J., 2005. Studies of reaction kinetics of methane hydrate dissociation in porous media. In: Proceedings of the Fifth International Conference on Gas Hydrates. Trondheim, Norway; June 2e6. http://www.icgh. org.

30. Nixon, J.F., 1986. The simulation of subsea saline permafrost. Can. J. Earth Sci. 23,

31. 2039e2046.

32. Ohgaki, K., Takano, K., Sangawa, H., Matsubara, T., Nakano, S., 1996. Methane exploitation by carbon dioxide from gas hydrates phase equilibria for CO2eCH4 mixed hydrate system. J. Chem. Eng. Jpn. 29 (3), 478e483.

33. Sloan, E.D., Koh, C.A., 2007. Clathrate Hydrates of Natural Gases, third ed. CRC Press, Boca Raton, FL, U.S.A.

34. Sun, Y.H., Lü, X.S., Guo, W., 2014. A review on simulation models for exploration and exploitation of natural gas hydrate. Arab. J. Geosci. 7 (6), 2199e2214.

35. Wang, P.K., Zhu, Y.H., Lu, Z.Q., Guo, X.W., Huang, X., 2011. Gas hydrate in the Qilian Mountain permafrost and its distribution characteristics. Geol. Bull. China 30 (12), 1839e1850.

36. Wang, P.K., Zhu, Y.H., Lu, Z.Q., Huang, X., Pang, S.J., Zhang, S., 2014. Gas hydrate stability zone migration occurred in the Qilian mountain permafrost, Qinghai, Northwest China: evidences from pyrite morphology and pyrite sulfur isotope. Cold Reg. Sci. Technol. 98, 8e17.

37. Wang, P.K., Zhu, Y.H., Lu, Z.Q., Huang, X., Pang, S.J., Zhang, S., Yang, K.L., 2013. Gas hydrate field identification methods in Qilian Mountain permafrost. Miner. Depos. 32 (5), 1045e1056.

38. Zhang, H.T., Zhang, H.Q., Zhu, Y.H., 2007. Gas hydrate investigation and research in China: present status and progress. Geol. China 34 (6), 953e961.

39. Zhao, J.F., Yu, T., Song, Y.C., Liu, D., Liu, W.G., Liu, Y., Yang, M.J., Ruan, X.K., Li, Y.H., 2013. Numerical simulation of gas production from hydrate deposits using a single vertical well by depressurization in the Qilian Mountain permafrost, Qinghai-Tibet Plateau, China. Energy 52, 308e319.

40. Zhou, Y.W., Guo, D.X., Qiu, G.D., Li, S.D., 2000. China Permafrost. Science Press, Beijing, China (in Chinese).

41. Zhu, Y.H., Liu, Y.L., Zhang, Y.Q., 2006. Formation condition of gas hydrates in permafrost of the Qilian Mountains, Northwest China. Geol. Bull. China 25 (1e2), 58e63.

42. Zhu, Y.H., Zhang, Y.Q., Wen, H.J., Lu, Z.Q., Jia, Z.H., Li, Y.H., Li, Q.H., Liu, C.L., Wang, P.K., Guo, X.W., 2010a. Gas hydrates in the Qilian mountain permafrost, Qinghai, Northwest China. Acta Geol. Sin. Engl. Ed. 84 (1), 1e10.

43. Zhu, Y.H., Zhang, Y.Q., Wen, H.J., Lu, Z.Q., Wang, P.K., 2010b. Gas hydrates in the Qilian Mountain permafrost and their basic characteristics. Acta Geosci. Sin. 31 (1), 7e16.

Chapter 7

Hydrodynamic Effects of Air Sparging on Hollow Fiber Membranes in a Bubble Column Reactor

Lijun Xia[a, b], Adrian Wing-Keung Law[a, b], and Anthony G. Fane[a, b]

[a]Singapore Membrane Technology Centre, 50 Nanyang Avenue, 639798 Singapore, Singapore
[b]School of Civil and Environmental Engineering, Nanyang Technological University, 50 Nanyang Avenue, 639798 Singapore, Singapore

ABSTRACT

Air sparging is now a standard approach to reduce concentration polarization and fouling of membrane modules in membrane bioreactors (MBRs). The hydrodynamic shear stresses, bubble-induced turbulence and cross flows scour the membrane surfaces and help reduce the deposit of foulants onto the membrane surface. However, the detailed quantitative knowledge on the effect of air sparging remains lacking in the literature due to the complex hydrodynamics generated by the gas–liquid flows. To date, there is no valid model that describes the relationship between the membrane fouling performance and the flow hydrodynamics. The present study aims to examine the impact of hydrodynamics induced by air sparging on the membrane fouling mitigation in a quantitative manner. A modelled hollow fiber module was placed in a cylindrical bubble column reactor at different axial heights with the trans-membrane pressure (TMP) monitored under constant flux conditions. The configuration of bubble column without the membrane module immersed was identical to that studied by Gan et al. (2011) using Phase Doppler Anemometry (PDA), to ensure a good quantitative understanding of turbulent flow conditions along the column height. The experimental results showed that the meandering flow regime which exhibits high flow instability at the 0.3 m is more beneficial to fouling alleviation compared with the steady flow circulation regime at the 0.6 m. The filtration tests also confirmed the existence of an optimal superficial air velocity beyond which a further increase is of no significant benefit on the membrane fouling reduction. In addition, the alternate aeration provided by two air stones mounted at the opposite end of the diameter of the bubble column was also studied to investigate the associated flow dynamics and its influence on the membrane filtration performance. It was found that with a proper switching interval and membrane module orientation, the membrane fouling can be effectively controlled with even smaller superficial air velocity than the optimal value provided by a single air stone. Finally, the testing results with both inorganic and organic feeds showed that the solid particle composition and particle size distribution all contribute to the cake formation in a membrane filtration system.

INTRODUCTION

Air sparging is now a standard approach to reduce concentration polarization and fouling of membrane modules in membrane bioreactors (MBRs). In the air sparged membrane system, shear stress is the critical parameter governing the foulant removal and enhancing the permeate flux. Shear stress is related to the viscosity and the velocity gradient at the membrane surface, which can be controlled by the hydrodynamic conditions of the reactor (Chan et al., 2007). However, the conditions are very complex due the gas–liquid two phase interactions, and they are also influenced by the membrane module configuration, gas sparger configuration as well as the reactor geometry (Meng et al., 2009).

Until recently, the focus on membrane fouling alleviation by air sparging was placed on the mean flow circulation in the reactor induced by bubbling. For example, Liu et al. (2000) studied the hydraulic characteristics in a submerged MBR system by measuring the average cross flow velocity distribution along the membrane surface. They concluded that a critical cross flow velocity of 0.3 m/s existed, below which the TMP increased sharply. Kang et al. (2008) developed an integrated numerical approach to quantify the hydrodynamics in both prototype and pilot MBR systems, with simulations targeting the mean flow circulation inside the MBRs. Khalili-Garakani et al. (2011) investigated the mechanisms that led to flux enhancement and fouling reduction by both experimental approaches and numerical simulations. They computed numerically the mean gas and liquid velocities using the k- turbulence closure, and suggested that the shear stresses on the membrane surface had a high correlation with the filtration resistance. These studies implied that the mean shear stress on the membrane is the key towards the reduction of fouling on the membrane surface.

In addition to the mean flow circulation, there is now an increasing recognition that turbulence fluctuations induced by bubbling also play an important role in membrane fouling control. Yeo et al. (2006) examined the filtration performance of submerged hollow fibers with and without bubbling, and quantified the relationship between the fouling performance and the unsteady flow conditions. His results clearly illustrated the importance of turbulent shear stresses on membrane fouling mitigation. Chan et al. (2011)found that the fouling rate for membranes subject to transient shear conditions was lower than for membranes subject to constant shear conditions. Various approaches have been proposed to enhance the fluid instability including turbulence promoters, pulsation, rotating membrane filter as well as secondary flows (Xu et al., 2002). In terms of aeration schemes, continuous injection has been typically adopted since the beginning of MBR development (Le-Clech et al., 2006). It can achieve good fouling control but the energy consumption is high. The realization of the importance of turbulence fluctuations to the membrane performance in MBRs has spawned a number of innovative aeration schemes for the air sparger design in recent years. They include the Eco-aeration scheme of Zenon›s units with alternate pulsing (Judd, 2011), as well as the Mempulse units by Siemens based on irregular pulsing (Siemens, 2009). Essentially, these recent schemes attempt to maximize the ability of fluctuating hydrodynamics in the reactor to achieve more effective fouling control.

The present study investigates the hydrodynamic effect of air sparging on the membrane performance in a lab scale cylindrical bubble column reactor. The relationship between the hydrodynamics induced by air sparging and the membrane fouling rate was explored quantitatively. Factors including the aeration strength (in terms of superficial air velocity), bubble-induced turbulent flow, air sparger geometry, membrane module orientation as well as the nature of feed solutions was examined. Typically, a total of twenty air flow rates from 0.1 to 2.0 L/min at 0.1 L/min interval were tested for each configuration, which corresponded to the superficial air velocity from 0.11 to 2.17 mm/s. The air flow rate was carefully varied so that the quantitative hydrodynamic conditions of the reactor could be controlled. The fouling performance was assessed by conducting the membrane filtration tests under constant flux conditions with the trans-membrane pressure (TMP) measured. The filtration tests were also conducted at different axial locations to study the effect of flow regimes on the cake formation. In addition, the alternate aeration provided by two air stones mounted at the opposite end of the diameter of the bubble column was also studied to investigate the associated flow dynamics and its influence on the membrane filtration performance. Factors such as the membrane orientation, switching interval as well as the nature of different feeds were also considered. In the following, we shall first describe the experimental setup and procedures before discussing the experimental results.

EXPERIMENTAL SETUP AND METHODOLOGY

Apparatus

Fig. 1 shows the schematic diagram of the experimental setup for the membrane filtration tests with air sparging. The setup consisted of a cylindrical bubble column reactor, a hollow fiber membrane module and a data logging system. The bubble column reactor had an inner diameter of 140 mm and an outer diameter of 152 mm, and the feed solution was typically filled to a depth of 1 m. A modelled hollow fiber membrane module was placed at both 0.3 and 0.6 m above the bottom of the reactor to examine the effect of the hydrodynamic conditions at these two locations. Constant flux conditions were maintained by carefully varying the rotational speed of a peristaltic pump (Spectral-Teknik (S) Pte Ltd) which was controlled by a programmable logic controller (PLC) system. The TMP changes of the membrane module were then monitored for 1 h, and raw data was logged and stored in a computer system for post-processing. During the experiment, a sensor was suspended at the top of the reactor with its tip touching the feed solution surface to regulate the feed level. It would send a signal to the PLC system to activate a feed pump if the feed level dropped too low. The air flow rate was determined by an air flow meter manufactured by Dwyer Company. A pulse dampener was installed to eliminate the pulsation of the pumping system.

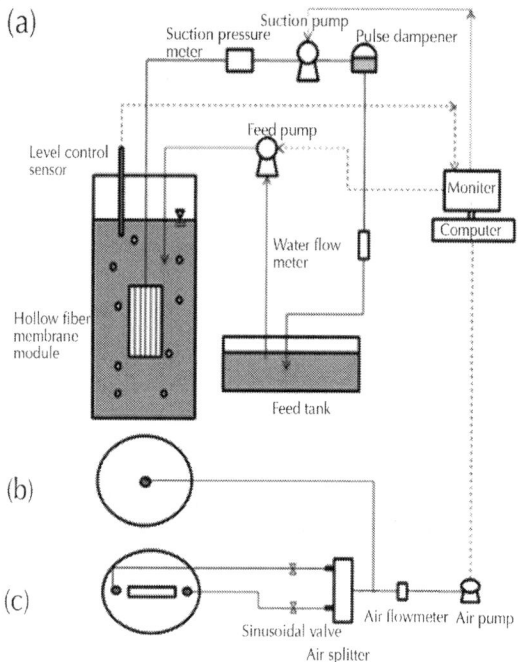

Figure 1: Experimental setup for the membrane filtration tests.

Fig. 1b and c show the central and alternate aeration schemes used in the membrane filtration tests, respectively. For the central aeration scheme, a perforated air stone with a diameter of 30 mm was mounted at the central bottom of the reactor for air bubble generation. This type of point air source is commonly used in the environmental industry for wastewater treatment due to its simplicity, high air flux and easy control (Liu and Tay, 2004; Gan et al., 2011). For the alternate aeration scheme, two air stones of 20 mm diameter each were placed at the opposite end of the diameter of the bubble column. Particular attention was paid towards the switching interval and its effect on membrane fouling. The curtain shape membrane module was aligned with as well as perpendicular to the alignment of the two air stones. For both locations, three alternate aeration patterns were tested with a switching interval of 0.5, 1 and 2 min. When the switching interval was specified, one of the air stones would release air for the period first, after which it would then be turned off and the other air stone was activated. The switching would repeat itself for 1 h. Thus, although the air sparging remained continuous, the hydrodynamics was very much different from the central aeration scheme. Fig. 2 shows a picture of bubbly flow at the 0.6 m location with the central aeration scheme and the superficial air velocity equal to 2.17 mm/s. It could be observed that the bubbles were dispersed in the continuous liquid phase. They varied widely in size but were typically near spherical. Note that only the homogeneous bubbly regime was considered in this study, and the experimental range did not include heterogeneous regimes with breakup or coalescence of the bubbles.

Figure 2: Picture of the bubbling flow with superficial air velocity of 2.17 mm/s at 0.6 m location.

Membrane Specifications

Hollow fiber membranes are used extensively in the water industry owing to several beneficial features: modest energy requirement, large surface area per unit volume, flexible, low operating cost and etc. Despite these advantages, they also have the disadvantage of high propensity to fouling due to poor hydrodynamics associated with the modules. To ensure that the hollow fiber membranes are used efficiently, it is important to design the module and its hydrodynamics in such a way that the fibers foul evenly in a manual manner.

Fig. 3 shows the hollow fiber membrane module used in the membrane filtration tests. The hollow fibers were polyvinylidene fluoride (PVDF) ultra-filtration membranes with 1.0 mm inner diameter and 2.0 mm outer diameter, and of around 0.1 μm pore size (MemStar Technology Ltd). The module consisted of two custom-made polyvinyl chloride (PVC) membrane holders and a detachable stainless steel membrane stand. Each fiber bundle was composed of twenty hollow fiber membranes of 10 cm effective length each. The hollow fibers were attached to a membrane support frame with the base end sealed with epoxy. High quality permeate was then extracted from the upper part of the module using a peristaltic pump.

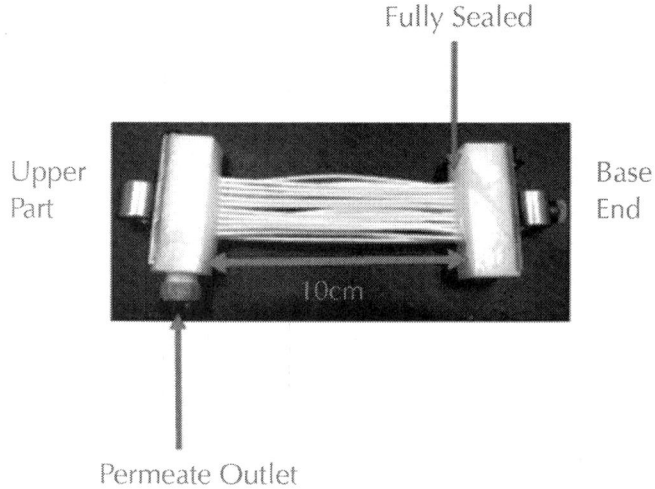

Figure 3: The hollow fiber membrane module used in the membrane filtration tests.

When multiple fiber bundles were used, a series of permeability tests were conducted to select the appropriate bundles for the experiments. The major objective of this permeability test was to increase the flux stepwise, and record the steady state value of the suction pressure which was normally 30 min after starting the measurements. The suction pressure versus flux curve was plotted, and the best fit line was obtained. The gradient of each line was then the permeability of the membrane module. The permeability values of different membrane modules were compared to make sure that the initial TMPs were close to each other. Those that showed large deviations of the initial TMP values were eliminated for use. Beyond this, the measurement of water permeability determined after backwashing whether the hollow fiber membranes could still be re-used or not.

When the filtration process ended, the hollow fiber membrane bundle was backwashed with permeate by reversing the rotating direction of the peristaltic pump for about 10 min. Subsequently, 1% of Terg-A-Zyme enzymatic detergent solution provided by Sigma–Aldrich was used to wash the module for another 10 min. After that, the module was rinsed with Milli-Q water to wash off the chemicals attached to the membrane surface, and finally the module was immersed in deionised water for further use. The measurement of water permeability after cleaning was necessary since the hollow fiber membrane would not be suitable for further use if its permeability could not be fully recovered.

It had been proven that there exists an optimum inner diameter for the hollow fibers for productivity maximization per unit energy, and a certain degree of fiber looseness and flexibility is very important (Cui et al., 2003). However, since the major focus in the present study is to correlate the hydrodynamics induced by air sparging with the membrane fouling performance, the fibers were thus glued tightly, i.e., 0% looseness, to eliminate the turbulent effect induced by the fiber movement.

Feed Solutions

Two kinds of feed solutions, i.e., the bentonite (Sigma–Aldrich) and yeast (Bake King's brand) solutions were tested since both of them are widely used as synthetic feed media for laboratory testing. Bentonite is an absorbent aluminium phyllosilicate. It is essentially impure clay consisting mostly of montmorillonite, which is a di-octahedral smectite mineral with a layered crystal structure with negative surface electrical charges arising from isomorphic substitutions of magnesium for aluminium in the crystal sheets. The clay swells in water by taking up inter-layer water, eventually leading to complete dispersion. The size of montmorillonite particles is thus not very well-defined, but they may be regarded as flat plate-like particles of lateral dimensions of the order of 1 μm and thickness of the order of 10 nm (Ní Mhurchú and Foley, 2008). Bentonite clay has been proven to have complex rheological characteristics, and the non-Newtonian nature of bentonite suspension can be expected to have a pronounced effect on the filtration characteristics (Benna et al., 1999; Güngör, 2000; Mahto and Sharma, 2004). Many studies have been conducted to investigate the fouling mechanisms using either dead- or cross-flow filtration with the bentonite suspension as the feed solution (e.g. Hamachi and Mietton-Peuchot, 1999; Benna et al., 2001).

As for yeast, it is a eukaryotic microorganism classified in the category of fungi. Yeast size is typically 3–4 μm in diameter but can vary greatly depending on the species. A notable feature of yeast is the large proportion of nitrogenous substances which it contains. The quantity varies a good deal according to the conditions of nutrition under which the yeast has been grown, but in general, more than one-half of the dry substances consist of proteins and other nitrogenous bodies. The proximate constituents other than nitrogen compounds include glycogen, gum, mucilage, fact, resinous matters and cellulose, together with a good proportion of mineral ingredients (Simmonds, 1919). Membrane filtration tests using yeast are of practical importance since yeast is one of the most important host of genetic modification for bio-product manufacturing (Sur and Cui, 2005).

In the experiments, the same mass concentration of 0.65 g/L was used for both feeds. The feed solution was prepared every other day with 10 g of bentonite or yeast powders weighted and mixed with water in a beaker. A stirring machine was used for mixing to assure a uniform dispersion of the particles. After that, the concentrated solution was poured into the reactor, and clean water was added to fill up the liquid height to 1 m. Since the bentonite particles swelled once mixed with water, the inorganic solution was usually prepared a day ahead to make sure that the majority of the solid particles no longer increased in size.

The dynamic viscosity of bentonite solution was also tested before the filtration experiments. Generally, the dynamic viscosity can be computed as a ratio of the shear stress over the shear rate. Fig. 4 shows the variations of dynamic viscosity for clean tap water and three different concentrations of bentonite solutions: 0.65, 1.3 and 2.6 g/L. The dynamic viscosities were found to be 8.83×10^{-4}, 9.27×10^{-4}, 9.33×10^{-4}, 9.80×10^{-4} correspondingly. Essentially the values were similar and the difference in viscosities between the clean water and the 0.65 g/L bentonite solution was only 2.57%. It could thus be expected that the flow conditions would also be similar in clean water and these low concentration bentonite solutions when subjected to air sparging.

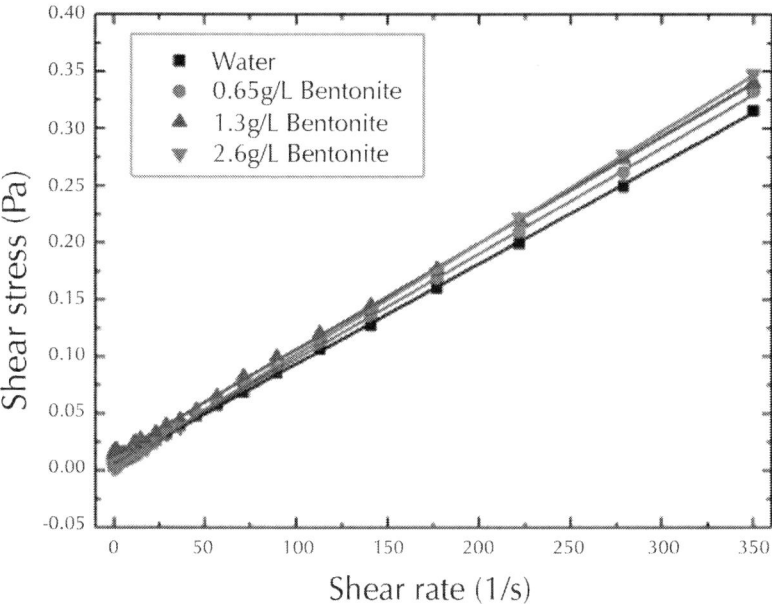

Figure 4: Variations of the dynamic viscosity for the clean tap water and three different concentrations of bentonite solutions of 0.65, 1.3 and 2.65 g/L.

Owing to the cylindrical geometry of the bubble column and the non-transparency nature of the feeds being tested in this study, the direct measurements of velocity distribution of the reactor with the feed solutions were difficult. However, since relatively low concentration of feeds was used, the flow pattern with the feed solutions can be assumed to be similar to the clean water. The bubble swarm could thus be studied through direct visual observation of the bubble movement in clear water. The extent of fouling control due to different flow conditions was mainly represented by the variation of membrane fouling rates.

Operating Conditions

Superficial Air Velocity

In the experiments, the hydrodynamics of the bubble column reactor was primarily controlled by the superficial air velocity. Although aeration demand is often given in terms of relative values such as air flow rate per membrane surface area (SADm) or air flow rate per permeate flow produced (SADp), the superficial air velocity is more appropriate to characterize the effect of hydrodynamics induced by air sparging for fouling control (Drews, 2010; Böhm et al., 2012). In the present study, a total of twenty air flow rates from 0.1 to 2.0 L/min at 0.1 L/min interval were examined for each configuration, corresponding to the superficial air velocity from 0.11 to 2.17 mm/s. The membrane performance was assessed at constant flux conditions by recording TMP values every 10 s for 1 h. The fouling rate dTMP/dt was then computed and compared. The detailed computation procedures are described in Section 3.1.1.

Bubble Column Reactor

As mentioned earlier, two locations of 0.3 and 0.6 m from the bottom of the reactor were investigated. These two locations were chosen due to the differences in hydrodynamic conditions. It had been observed that with the central aeration scheme, the hydrodynamics would vary along the height of the bubble column as illustrated by the instantaneous and time-averaged flow field sketched in Fig. 5. Gan et al. (2011)measured the two phase hydrodynamics using the Phase Doppler Anemometry (PDA) technique using the same bubble column geometry with clean water as the feed solution. They found that in the region from the tip of the air stone z = 50 mm to about z = 300 mm, the injected bubble swarm rose from the air sparger forming a bubble plume. Meandering flows were dominant and the instantaneous circulation cells were not symmetrical about the centreline. Towards the middle portion of the reactor, the flow became more orderly. Near the surface, the bubbly flow was more or less uniform, and bubbles were well distributed across the entire cross-section (see Fig. 5(a)). For the time average behaviour of the flow pattern as shown in Fig.

5(b), the bubble plume was developing along the z-axis and in the middle portion, the bubbles were concentrating in the central core and the flow became relatively uniform. Near the air sparger, large scale circulation existed due to the wandering of the bubble plume and hence much less bubbles were concentrated except for some tiny ones. In this regard, the two locations were picked with 0.3 m located in the meandering region and 0.6 m in the transition region. Due to the difference in the hydrodynamic behaviours at these two locations, the effects of flow regime on the membrane filtration performance can thus be characterized and optimized for fouling minimization.

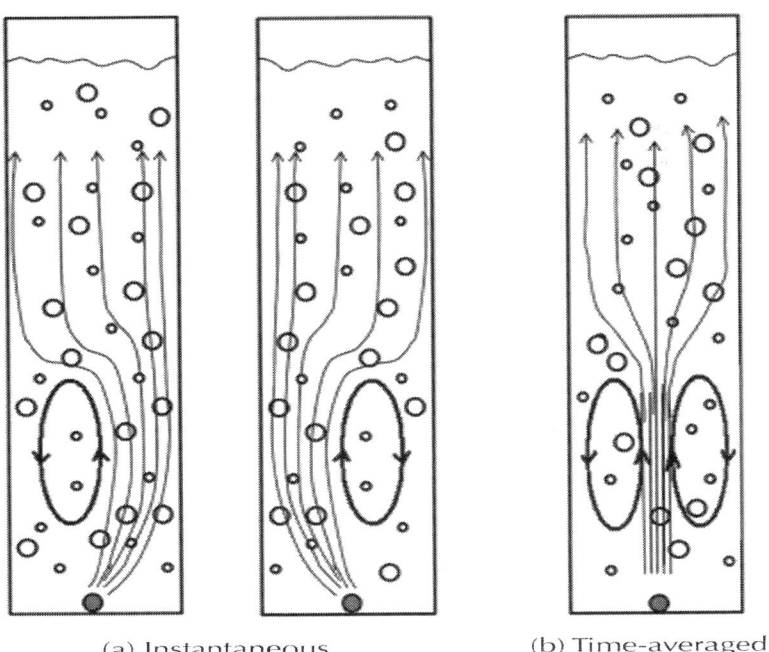

(a) Instantaneous   (b) Time-averaged

Figure 5: Sketch of the flow field by central aeration in the bubble column (from Gan et al., 2011).

Aeration Schemes

The energy cost for aeration is an important consideration for the MBR operation (Buer and Cumin, 2010). Thus, there has been intense interest to optimize the aeration configuration to achieve energy minimization on membrane fouling control. The issue is however intricate, for example, deriving the optimal bubble size for a specific MBR is not straightforward (Fane et al., 2005). Altering the flow regimes at the membrane surface towards a more turbulent flow pattern is normally beneficial for foulant removal (Flemming, 1995), but the level of turbulence depends critically on the reactor configuration. In terms of aeration schemes, continuous injection is typically adopted in the beginning of MBR development. It achieves good fouling control, but the energy consumption can be high. Since then, many innovative schemes have been developed with good results. These include the alternate aeration scheme which bubbling is switched alternately between the two sides of the membrane module. It was proven that alternating the injection of air was highly beneficial for the overall filtration system performance (Guibert et al., 2002). Hence, in the present study, in addition to deploying a single air stone to provide continuous bubbling, a two air stone configuration was also employed to generate the alternate bubbling pattern. Both the central and alternate aeration schemes are commercially applied in the industry at present, and the two air stone configurations were adopted with the intent to mimic the basic switching principle.

Fig. 6 depicts the transient flow conditions in the reactor when the alternate aeration scheme was applied. At the 0.3 m location, the bubble plume was developing spatially but the plume adhered to the side of the membrane module due to the near wall placement of the two air stones. Large flow circulation could be observed near the air sparger region. Moving up to the 0.6 m location, the bubble plume became developed with bubbles dispersed throughout the cross-section of the reactor. This would continue for the switching interval before the bubbly air stone was changed to the opposite one. Same testing locations, namely, the 0.3 and 0.6 m to the bottom of the reactor were chosen since the 0.3 m was located in a region with big flow circulation while the 0.6 m location was located at the place where the flow behaved more regular.

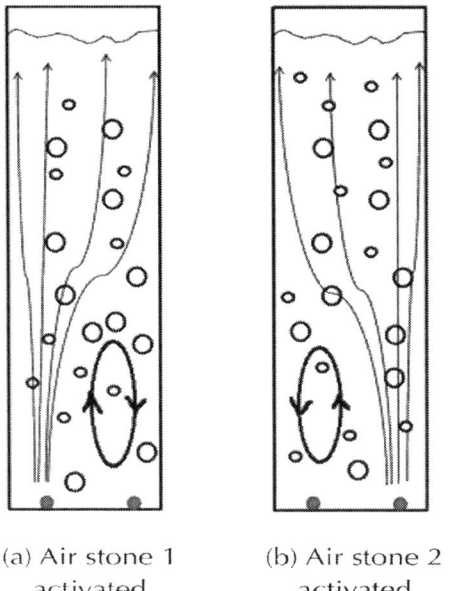

(a) Air stone 1
activated

(b) Air stone 2
activated

Figure 6: Transient flow condition induced by alternate aeration in the bubble column.

Permeate Flux

The type of fouling considered in this filtration experiments was primarily surface fouling. Hence our main interest is on the hydrodynamic conditions in the vicinity to the membrane surface caused by bubble movement, and the detailed deposition mechanisms in the membrane surface were not considered. Continuous filtration was carried out within a relatively short operating time ($\sim$1.0 h), and a permeate flux of 71.6 L/m^{2}hr was adopted in the experiments to produce the cake layer. During the filtration process, the highest TMP did not exceed 60 kpa, which was within the maximum operating pressure recommended by the manufacturer for hollow fiber membranes.

Membrane Module Orientation

When the alternate aeration scheme was utilized in the membrane filtration experiments, two different module orientations were tested:

alignment with the two air stones (0°) and perpendicular to the alignment (90°) (see Fig. 7). The objective was to investigate the effect of the direction of cross flow to the membrane curtain on the fouling performance. In this case, the 1.0 min switching interval was examined.

Membrane Module

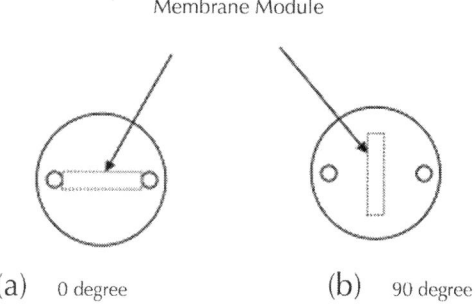

(a) 0 degree (b) 90 degree

Figure 7: Membrane module orientations: (a) 0° and (b) 90°.

RESULTS AND DISCUSSION

Measurements and Calculations

Trans-Membrane Pressure (TMP) and Fouling Rate

Generally, the filtration flux and TMP are the best indicators of membrane fouling. The TMP is calculated as follows:

$$P_{TMP} = \frac{(P_f + P_r)}{2} - P_p$$

(1)

where P_{TMP} is the trans-membrane pressure (TMP), P_f is the feed pressure, P_r is the retentate pressure and P_p is the permeate pressure. In the present study, the TMP values were measured under constant flux conditions and the temporal rise in TMP was used to assess the fouling rate of the system as follow:

$$\frac{dTMP}{dt} = \frac{dP_s}{dt}$$

(2)

where P_s is the suction pressure. In the data analysis, the fouling rates were compared rather than the absolute values of TMP themselves.

Superficial Air Velocity

The superficial air velocity, which is a very important measure of the system performance, is equal to the volumetric flow rate divided by the cross-sectional flow area. In the experiments, a cylindrical reactor was used and thus the average superficial air velocity could be computed by the following equation:

$$V_a = \frac{Q}{A}$$

(3)

where V_a is the average superficial air velocity, Q is the air flow rate and A is the cross-sectional area of the cylindrical reactor.

Continuous Aeration

The Effect of Flow Regime

We first examined the effect of superficial air velocity on the membrane filtration performance with 0.65 g/L concentration of bentonite solution using the central aeration scheme. Fig. 8 illustrates the development of fouling rate at the constant flux of 71.6 L/m²hr with various superficial air velocities at the 0.3 and 0.6 m locations. Firstly, we compared the membrane performance at the two locations with the increase of superficial air velocity. Initially, the fouling rate at the 0.3 m location (~15.84 kpa/hr) was higher than that at the 0.6 m location (~12.24 kpa/hr) when the superficial air velocity was around 0.1 mm/s. Along with the increase of aeration supply, lower fouling rate was found at the 0.3 m location compared with the 0.6 m location. The quantitative results suggested that the meandering flow regime dominant at the 0.3 m location was more beneficial to fouling alleviation. Comparing with a relatively steady circulation pattern at the 0.6 m location, the unsteady flow condition at the 0.3 m location was more effective in eroding the deposition layer. Hence for a specific MBR system, the hydrodynamics should be characterized and the effect maximized such as by immersing the membrane modules in the highly turbulent region to minimize fouling.

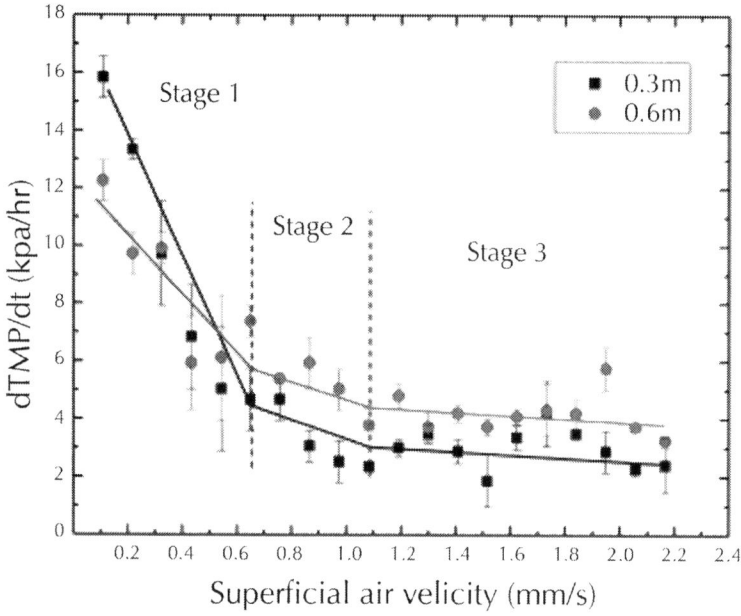

Figure 8: Fouling rates for 0.65 g/L bentonite solution at 0.3 and 0.6 m locations.

The Effect of Superficial Air Velocity

Next, we studied the variations of fouling rates with different superficial air velocities. It could be observed from Fig. 8 that with the increase of superficial air velocity, the effectiveness of bubbling towards membrane fouling alleviation could be categorized into three stages. The threshold of each stage could be distinguished by the change of the gradient as indicated in Fig. 8. When the superficial air velocity was small and below 0.65 mm/s, there was a fast reduction stage in which the fouling rate decreased substantially when the superficial air velocity just increased slightly. This was then followed by a slow mitigation stage from 0.65 to 1.1 mm/s, whereby the fouling rate continued to decrease but at a much slower pace compared to the first stage. When the superficial air velocity was increased beyond 1.1 mm/s, however, a limiting stage was reached and no further reduction of the membrane fouling rate could be observed. The above quantitative results suggested that introducing a small amount of air to the membrane filtration system could improve the membrane performance to a great extent, as illustrated in the first stage (Fig. 8). This is also well recognized in the industry, and explains the wide popularity of using air sparging in MBRs for membrane fouling control. Additionally, the results confirmed the existence of an optimal superficial air velocity for the air sparging of a membrane system beyond which further increases have no benefit on fouling mitigation. Many earlier researches had also pointed to the existence of this optimal aeration value (e.g. Ueda et al., 1997; Sofia et al., 2004; Le-Clech et al., 2006). Clearly, the determination of this optimal superficial air velocity for a specific configuration of membrane filtration system is of great importance in order that the energy cost for the air sparging can be optimised.

Alternate Aeration

The Effect of Flow Regime

Fig. 9 shows the variations of fouling rate at 0.3 and 0.6 m locations when the three switching intervals T of 0.5, 1.0 and 2.0 min were employed for 0.65 g/L bentonite solution. First of all, the effects of flow regime at the two locations were compared. With T = 0.5 min, the 0.6 m location showed a better filtration performance when the superficial air velocity was higher than 0.65 mm/s. When T was increased to 1.0 min, the variations of fouling rate for the two locations became more or less similar. A local minimum was found for both locations and the magnitudes were about the same as the asymptotic minimum fouling rate. When the switching interval was further extended to T = 2.0 min, the two locations showed substantial fouling reduction with the increase of superficial air velocity and lower fouling rate was found at the 0.6 m location.

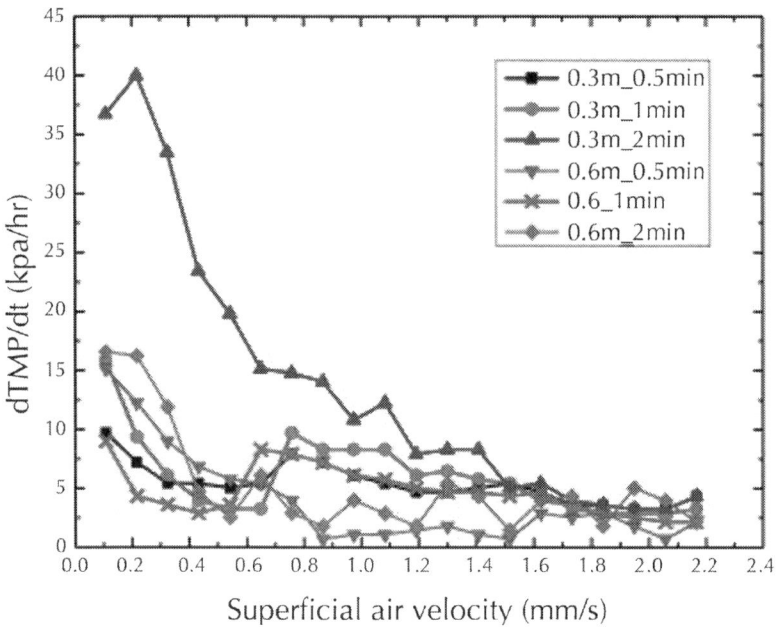

Figure 9: The variations of TMPs with alternate aeration at 0.5, 1 and 2 min switching interval at 0.3 and 0.6 m locations for 0.65 g/L bentonite solution.

The quantitative results suggested that for either shorter (0.5 min) or longer (2.0 min) switching interval, 0.6 m was a more desirable location to place the membrane module due to lower fouling rates. However, the occurrence of local minimum for both locations when 1.0 min was employed implied that a very small superficial air velocity was able to achieve an effective fouling control. Under this circumstance, where the membrane module was placed became less critical since either location could provide a good hydrodynamic condition for fouling control. In this regard, the aeration demand can be significantly reduced. The current results also demonstrated in quantitative details the complexity of the two phase hydrodynamics of the bubble column reactor.

The Effect of Superficial Air Velocity

The membrane filtration performance was also studied with the increase of superficial air velocity. FromFig. 9, it is obvious that three stages can also be distinguished for the fouling performance with the increase of superficial air velocity at the 0.3 m location, namely, the fast reduction, slow mitigation and limiting stages. Generally, 0.5 and 1.0 min alternate aerations showed similar decreasing trend as the central aeration in terms of fouling rate, whereas the fouling rates were much higher for the 2.0 min alternate aeration when small superficial air velocity (<0.65 mm/s) was applied. Between 0.65 and 1.50 mm/s, the fouling rate for the three switching intervals decreased with the increase of superficial air velocity but at a relatively slower pace. Beyond 1.50 mm/s, all three switching intervals exhibited similar fouling rate trend, i.e., towards an asymptotic minimum fouling rate. The experimental results again confirmed that an optimal superficial air velocity existed for the alternate aeration scheme similar to the central aeration scheme. The optimal superficial air velocity was roughly about the same, namely, 1.50 mm/s for all the three switching intervals.

For the 0.6 m location, generally, all the three switching intervals showed relatively lower fouling rates with the increase of superficial air velocity compared with the 0.3 m location. Among them, the 2.0 min switching interval had lower fouling rates compared with the 0.3 m location when small superficial air velocity was applied (<0.55 mm/s). Moreover, the 0.5 min switching interval achieved a local minimum fouling rate between 0.8 and 1.5 mm/s for both locations. The results suggested that besides the magnitude of superficial air velocity which was identical for the two locations, the local hydrodynamics due to the alternate aeration also played a significant role on the membrane fouling performance. The occurrence of the local minimum for both locations again demonstrated the complexity of the hydrodynamic conditions inside the reactor driven by alternate aeration.

The Effect of Switching Interval

From Fig. 9, the fouling rates were much higher for the 2.0 min alternate aeration when small superficial air velocity (<0.65 mm/s) was applied. The large fouling rate obtained for the 2.0 min alternate aeration showed that the switching interval cannot be chosen randomly and must be carefully determined. In general, the curves for 0.5 and 1.0 min showed similar trends but the magnitude of the local minimum was smaller for 1.0 min switching interval. Hence at the 0.3 m location, 1.0 min switching interval is more effective in eroding the cake layer compared with the other two switching intervals.

For the 0.6 m location, all the three switching intervals showed relatively lower fouling rates with the increase of superficial air velocity compared with the 0.3 m location. The superiority of 1.0 min switching interval no longer existed but the local minimum at 1.0 min still served as the desirable operating condition for better fouling control due to low fouling rates and small aeration demand.

From Fig. 9, it is clear that the flow regime, superficial air velocity as well as the switching interval all plays a role in influencing the hydrodynamics of the bubble column reactor when alternate aeration scheme was employed. Although the flow behaviour became much more complex with alternate aeration, it was found that there was an optimal superficial air velocity existed for both aeration schemes. The magnitude of this optimal superficial air velocity might change when operating condition changes, however, the three stage membrane filtration performance with the increase of superficial air velocity is always applicable for a membrane filtration system.

The Effect of Membrane Module Orientation

Fig. 10 illustrates the variations of membrane fouling rates for the two module orientations with the superficial air velocity changing from 0.11 to 1.1 mm/s. From the figure, it is obvious that the fouling rate decreased for both orientations with the increase of superficial air velocity. However, the fouling rate was always the highest for the 90° orientation at the 0.3 m location. This was followed by the 90° orientation at the 0.6 m location which at first the fouling rate was also high but then it quickly dropped to a relatively low level. The results suggested that the 0° was more beneficial compared with the 90° orientation due to lower fouling rates. This might be attributed to the fact that only one side of the membrane was scoured at any one time for the 90° orientation, while the other side could therefore get fouled quickly. For the 0° orientation, however, both sides of the module could still be cleaned at the same time despite the fact that the scouring action would be stronger for the part of the membrane modules closer to the active air stone.

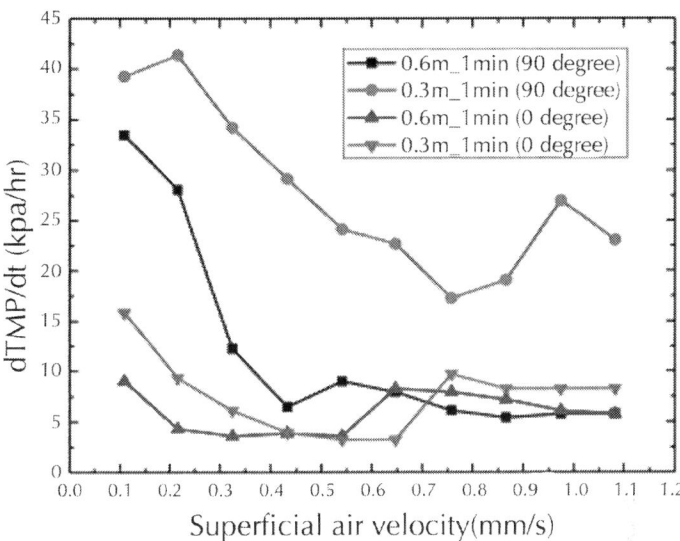

Figure 10: Variations of TMPs at the two membrane orientations in 0.65 g/L bentonite solution.

From Fig. 10, it can also be observed that for the 90° orientation, the 0.6 m location showed a better membrane performance compared with the 0.3 m location. This phenomenon can be correlated to the development of the bubble plume. Due to the immersion of the membrane module, the mixing inside the reactor was reduced. At 0.3 m, the alternate aeration was only able to reach part of the membrane surface whereas moving up to the 0.6 m location, the bubbling plume was more developed and could expand to a larger area, which as a result promoted the mixing and helped control fouling. Beyond 0.75 mm/s, the membrane performance at the 0.6 m location for the 90° orientation was almost identical to the 0° orientation. Due to general better membrane performance achieved by the 0° orientation, the 0° orientation was chosen to study the effect of switching interval, further with the results already discussed in the previous section.

The Effect of Feed Solution

Two types of feeds, namely, bentonite and yeast solution were examined in the current study. Fig. 11 illustrates the variations of fouling rates against the superficial air velocity for both the bentonite and yeast solutions with the same mass concentration of 0.65 g/L. The operating conditions with the yeast solutions were identical to those presented in Fig. 8 for the bentonite solutions, except that the constant flux was reduced from 71.6 L/m²hr to 28.6 L/m²hr. The reduction of permeate flux was necessary as otherwise the foulant deposited too fast with the yeast solution. Further explanation of using lower flux is provided in the following. From the figure, it could be observed that initially the fouling rate for the yeast solution was much higher compared with the bentonite solution. With the increase of superficial air velocity, the pattern of the three stages described above, namely fast reduction, slow mitigation and limiting stages, appeared. When 0.65 mm/s superficial air velocity was reached, the fouling rates no longer decreased for both the 0.3 and 0.6 m location, and a minimum fouling rate was achieved. Hence, the optimal superficial air velocity would be 0.65 mm/s for the yeast solution in this case. The experimental results illustrated that the action of hydrodynamics on membrane fouling performance was similar for both feeds. The smaller magnitude of the minimum fouling rate obtained with the yeast solution actually implied that air sparging might be practically more favourable to remove bio-particles rather than inorganic feeds.

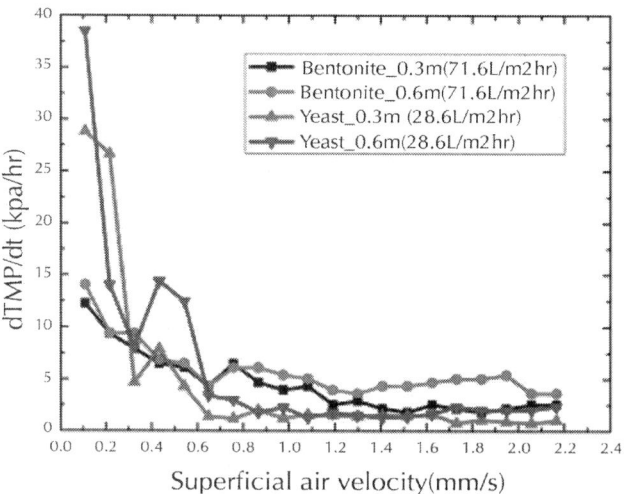

Figure 11: Fouling rates for 0.65 g/L bentonite and yeast solutions at 0.3 and 0.6 m locations.

Regarding the filtration flux, we originally conducted the membrane filtration tests at 71.6 L/m²hr with the yeast solution as well. The experimental results are shown in Fig. 12(a) and (b). In the two figures, the TMP values shot up to 100 kpa for both locations as soon as the filtration was initiated. The fouling rate was thus much higher than the bentonite solution showed in Fig. 8 with the same mass concentration. This could be partially attributed to the cake formation mechanisms which were considerably different between bentonite and yeast solutions. Another reason might be due to the presence of supernatant in the yeast solution which contained cell debris and extracellular polymer substances (EPS), with EPS already being proven to be the major factor that causes rapid foulant accumulation (Sur and Cui, 2005). In addition, the difference in fouling rate could be contributed to the difference of particle numbers in the two feeds. The particle size distributions were measured for both solutions using Mastersizer 2000. It was determined that the bentonite particles had an average diameter of 5.83 μm, while for the yeast particles the average diameter was about 4.95 μm. Although the diameters were similar, the densities were vastly different for the two solutions. After calculation, it was found out that the particles in the yeast solution were ~3.5 times the number of particles in the bentonite solution. Last but not the least, the presence of very small size particles in the yeast solution might also led to potential pore blockage in the membranes. Hence, under the same operating conditions, the membrane was more prone to fouling in the yeast solution due to the existence of supernatant, the much larger number of particles as well as the presence of extremely small size particles. As a result, the membrane fouling behaviour could only be monitored if the trans-membrane flux was reduced. After some trial and errors, the flux in the yeast solution was finalized to 28.6 L/m²hr.

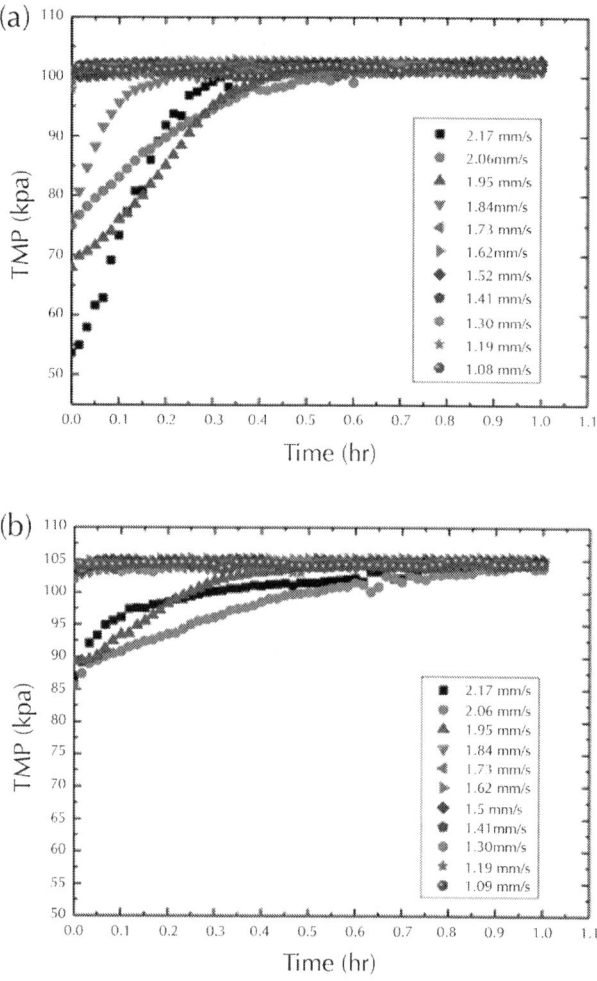

Figure 12: Variations of TMPs at (a) 0.3 m and (b) 0.6 m locations for 0.65 g/L yeast solution.

CONCLUSIONS

The present study confirms the existence of an optimal superficial air velocity for the air sparging of a membrane module. The optimal superficial air velocity would serve as the upper limit for the assessment of the superficial air velocity associated with the sustainable flux. The experimental results also showed that the meandering flow regime which exhibits higher flow instability is more effective towards fouling alleviation.

Two different air sparger configurations, namely, the central and alternate aeration schemes, were tested. Various factors such as the superficial air velocity, flow regime, switching interval and the membrane module orientation were investigated. Compared with the central aeration scheme, the alternate aeration produced by the two air stones with 1.0 min switching interval can greatly reduce the aeration demand by applying the superficial air velocity at the local minimum with the 0° membrane module orientation. In addition, the application of inorganic and organic feeds for membrane filtration demonstrated that the particle composition as well as particle size distribution all contributes to the cake formation on the membrane surface.

ACKNOWLEDGEMENT

The authors would like to express their grateful gratitude to the Singapore National Research Foundation under its Environmental and Water Technologies Strategic Research Programme and administered by the Environmental and Water Industry Programme Office (EWI) of PUB, and Singapore Membrane Technology Centre and Nanyang Technological University for providing the equipment and space to this research project.

REFERENCES

1. Benna, M., Kbir-Ariguib, N., Clinard, C., Bergaya, F., 2001. Static filtration of purified sodium bentonite clay suspensions. Effect of clay content. Applied Clay Science 19 (1e6), 103e120.

2. Benna, M., Kbir-Ariguib, N., Magnin, A., Bergaya, F., 1999. Effect of pH on rheological properties of purified sodium bentonite suspensions. Journal of Colloid and Interface Science 218 (2), 442e455.

3. Bo"hm, L., Drews, A., Prieske, H., Be´ rube´, P.R., Kraume, M., 2012. The importance of fluid dynamics for MBR fouling mitigation. Bioresource Technology 122 (0), 50e61.

4. Buer, T., Cumin, J., 2010. MBR module design and operation. Desalination 250 (3), 1073e1077.

5. Chan, C.C.V., Be´ rube´, P.R., Hall, E.R., 2007. Shear profiles inside gas sparged submerged hollow fiber membrane modules. Journal of Membrane Science 297 (1e2), 104e120.

6. Chan, C.C.V., Be´ rube´, P.R., Hall, E.R., 2011. Relationship between types of surface shear stress profiles and membrane fouling. Water Research 45 (19), 6403e6416.

7. Cui, Z.F., Chang, S., Fane, A.G., 2003. The use of gas bubbling to enhance membrane processes. Journal of Membrane Science 221 (1e2), 1e35.

8. Drews, A., 2010. Membrane fouling in membrane bioreactors e characterisation, contradictions, cause and cures. Journal of Membrane Science 363 (1e2), 1e28.

9. Fane, A.G., Yeo, A., Law, A., Parameshwaran, K., Wicaksana, F., Chen, V., 2005. Low pressure membrane processes e doing more with less energy. Desalination 185 (1e3), 159e165.

10. Flemming, H.C., 1995. Biofouling bei Membranprozenssen. Springer-Verlag, Berlin, Heidelberg.

11. Gan, Z.W., Yu, S.C.M., Law, A.W.K., 2011. Hydrodynamic stability of a bubble column with a bottom-mounted point air source. Chemical Engineering Science 66 (21), 5338e5356.

12. Guibert, D., Aim, R.B., Rabie, H., Coˆ te´, P., 2002. Aeration performance of immersed hollow-fiber membranes in a bentonite suspension. Desalination 148 (1e3), 395e400.

13. Gu¨ngo¨ r, N., 2000. Effect of the adsorption of surfactants on the rheology of Na-bentonite slurries. Journal of Applied Polymer Science 75 (1), 107e110.

14. Hamachi, M., Mietton-Peuchot, M., 1999. Experimental investigations of cake characteristics in crossflow microfiltration. Chemical Engineering Science 54 (18), 4023e4030.

15. Judd, S., 2011. Chapter 3 e Design, Operation and Maintenance, second ed.. In: The MBR Book Butterworth-Heinemann, Oxford, pp. 209e288.

16. Kang, C.-W., Hua, J., Lou, J., Liu, W., Jordan, E., 2008. Bridging the gap between membrane bio-reactor (MBR) pilot and plant studies. Journal of Membrane Science 325 (2), 861e871.

17. Khalili-Garakani, A., Mehrnia, M.R., Mostoufi, N., Sarrafzadeh, M.H., 2011. Analyze and control fouling in an airlift membrane bioreactor: CFD simulation and experimental studies. Process Biochemistry 46 (5), 1138e1145.

18. Le-Clech, P., Chen, V., Fane, T.A.G., 2006. Fouling in membrane bioreactors used in wastewater treatment. Journal of Membrane Science 284 (1e2), 17e53.

19. Liu, R., Huang, X., Wang, C., Chen, L., Qian, Y., 2000. Study on hydraulic characteristics in a submerged membrane bioreactor process. Process Biochemistry 36 (3), 249e254.

20. Liu, Y., Tay, J.-H., 2004. State of the art of biogranulation technology for wastewater treatment. Biotechnology Advances 22 (7), 533e563.

21. Mahto, V., Sharma, V.P., 2004. Rheological study of a water based oil well drilling fluid. Journal of Petroleum Science and Engineering 45 (1e2), 123e128.

22. Meng, F., Chae, S.-R., Drews, A., Kraume, M., Shin, H.-S., Yang, F., 2009. Recent advances in membrane bioreactors (MBRs): membrane fouling and membrane material. Water Research 43 (6), 1489e1512.

23. Nı́ Mhurchú , J., Foley, G., 2008. Dead-end and Crossflow Microfiltration of Yeast and Bentonite Suspensions: Experimental and Modelling Studies Incorporating the Use of Artificial Neural Networks. School of Biotechnology Dublin City University, p. 287.

24. Siemens, 2009. MemPulse Membrane Bioreactor (MBR) System. from. http://www.water.siemens.com/.

25. Simmonds, C., 1919. Composition of Yeast. Alcohol, Its Production, Properties, Chemistry, and Industrial Applications. Macmillan And Co.

26. Sofia, A., Ng, W.J., Ong, S.L., 2004. Engineering design approaches for minimum fouling in submerged MBR. Desalination 160 (1), 67e74.

27. Sur, H.W., Cui, Z.F., 2005. Enhancement of microfiltration of yeast suspensions using gas sparging e effect of feed conditions. Separation and Purification Technology 41 (3), 313e319.

28. Ueda, T., Hata, K., Kikuota, Y., Seino, O., 1997. Effects of aeration on suction pressure in a submerged membrane bioreactor. Water Research 31, 489e494.

29. Xu, N., Xing, W., Xu, N., Shi, J., 2002. Application of turbulence promoters in ceramic membrane bioreactor used for municipal wastewater reclamation. Journal of Membrane Science 210 (2), 307e313.

30. Yeo, A.P.S., Law, A.W.K., Fane, A.G., 2006. Factors affecting the performance of a submerged hollow fiber bundle. Journal of Membrane Science 280 (1e2), 969e982.

Citations

CHAPTER 1

Vijay H Honkalaskar, Upendra V Bhandarkar, and Milind Sohoni, Development of a Fuel Efficient Cookstove through a Participatory Bottom-up Approach, doi:10.1186/2192-0567-3-16.

CHAPTER 2

Jiaxin Sun, Fulong Ning, Nengyou Wu, Shi Li, Ke Zhang, Ling Zhang, Guosheng Jiang, and V.F. Chikhotkin, The Effect of Drilling Mud Properties on Shallow Lateral Resistivity Logging of Gas Hydrate Bearing Sediments, doi:10.1016/j.petrol.2014.12.015.

CHAPTER 3

J.Y. Lee, G.-Y. Kim, N.K. Kang, B.-Y. Yi, J.W. Jung, J.-H. Im, B.-K. Son, J.-J. Bahk, J.-H. Chun, B.-J. Ryu, D.S. Kim, Physical properties of sediments from the Ulleung Basin, East Sea: Results from Second Ulleung Basin Gas Hydrate Drilling Expedition, East Sea (Korea), Marine and Petroleum Geology, Volume 47, November 2013, Pages 43-55, ISSN 0264-8172, http://dx.doi.org/10.1016/j.marpetgeo.2013.05.017.

CHAPTER 4

Ry Bloomdahl, Noura Abualfaraj, Mira Olson, and Patrick L. Gurian, Assessing Worker Exposure to Inhaled Volatile Organic Compounds from Marcellus Shale Flowback Pits, Doi:10.1016/j.jngse.2014.08.018.

CHAPTER 5

Prabhansu, Malay Kr. Karmakar, Prakash Chandra, Pradip Kr. Chatterjee, A review on the fuel gas cleaning technologies in gasification process, Journal of Environmental Chemical Engineering, Available online 16 February 2015, ISSN 2213-3437, http://dx.doi.org/10.1016/j.jece.2015.02.011.

CHAPTER 6

Youhong Sun, Bing Li, Wei Guo, Xiaoshu Lü, Yongqin Zhang, Kuan Li, Pingkang Wang, Guangrong Jin, Rui Jia, and Lili Qu, Comparative analysis of the Production trial and Numerical simulations of Gas Production from Multilayer Hydrate Deposits in the Qilian Mountain permafrost, Journal of Natural Gas Science and Engineering, Volume 21, November 2014, Pages 456-466, ISSN 1875-5100, http://dx.doi.org/10.1016/j.jngse.2014.09.005.

CHAPTER 7

Lijun Xia, Adrian Wing-Keung Law, Anthony G. Fane, Hydrodynamic effects of air sparging on hollow fiber membranes in a bubble column reactor, Water Research, Volume 47, Issue 11, 1 July 2013, Pages 3762-3772, ISSN 0043-1354, http://dx.doi.org/10.1016/j.watres.2013.04.042.

Index